Techniques of high resolution nuclear magnetic resonance spectroscopy

Techniques of high resolution nuclear magnetic resonance spectroscopy

W. McFarlane and R. F. M. White

London   Butterworths

THE BUTTERWORTH GROUP

ENGLAND
Butterworth & Co (Publishers) Ltd
London: 88 Kingsway, WC2B 6AB

AUSTRALIA
Butterworths Pty Ltd
Sydney: 586 Pacific Highway, NSW 2067
Melbourne: 343 Little Collins Street, 3000
Brisbane: 240 Queen Street, 4000

CANADA
Butterworth & Co (Canada) Ltd
Toronto: 14 Curity Avenue, 374

NEW ZEALAND
Butterworths of New Zealand Ltd
Wellington: 26–28 Waring Taylor Street, 1

SOUTH AFRICA
Butterworth & Co (South Africa) (Pty) Ltd
Durban: 152–154 Gale Street

First published 1972

ISBN 0 408 70414 4

Filmset and printed Offset Litho in Great Britain
by Cox & Wyman Ltd, London, Fakenham and Reading

Contents

1 Introduction

The first successful n.m.r. experiments, carried out independently at two universities in the U.S.A., were reported in 1946. The technique was developed by physicists and, initially, applications were concerned with measurements of properties such as magnetogyric ratios of nuclei, nuclear spin relaxation times, and internuclear distances in solids. From 1949 evidence accumulated to show that, for a particular nuclear species in a given magnetic field, the resonance frequency depends on the chemical environment of the nucleus. Discovery of this effect, now known as the chemical shift, led to the use of n.m.r. as a technique in structural chemistry and this in turn encouraged the development of high resolution methods. During this evolution it was discovered that liquid samples gave spectra showing fine structure due to spin-spin interaction; also the effects of chemical exchange on the appearance of spectra came to be appreciated.

In addition to becoming important in research, high resolution n.m.r. has had a considerable influence on the teaching of chemistry. It was probably one of the main factors encouraging lecturers to introduce spectroscopic techniques at an early stage in the education of a chemist. Many undergraduates become acquainted with n.m.r. in the first year of their degree course. To these students it is natural to try to distinguish between *normal-* and *iso-* propyl groups from the number of peaks in ^{1}H n.m.r. spectra. When dealing with acetylacetone, the student learns that its n.m.r. spectrum shows peaks arising from both the keto and the enol forms (Figure 1.1), the relative proportions of the two tautomers being determinable from the relative areas of appropriate peaks. At a slightly more advanced stage, the student may come to distinguish primary, secondary, and tertiary alcohols by means of the spin-spin splitting pattern appearing in the peak due to the —OH group (Figure 1.2), while *cis-* and *trans-* isomers of ethylenic compounds may be differentiated by the magnitude of the H–H coupling constants across the double bond. The uses of n.m.r. are not restricted to the study of organic compounds; increasingly it is used in the investigation of inorganic materials and the examination of kinetic processes.

As this book is a practical manual, only a brief and simplified account will be given of theoretical principles, and more detailed treatments can be found in the standard texts mentioned below. The books by Abragam[1] and Andrew[2] are advanced monographs dealing with the principles and methods

1

2

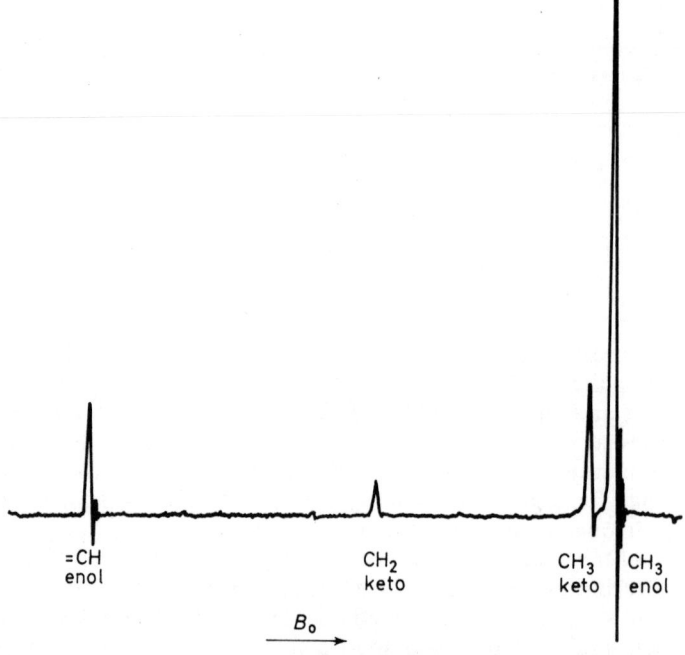

=CH
enol

CH$_2$
keto

B_o

CH$_3$
keto

CH$_3$
enol

Figure 1.1. The ^{1}H n.m.r. spectrum of acetylacetone. The peak due to the proton of the —OH group in the enol tautomer occurs at low field and is not shown in this spectrum

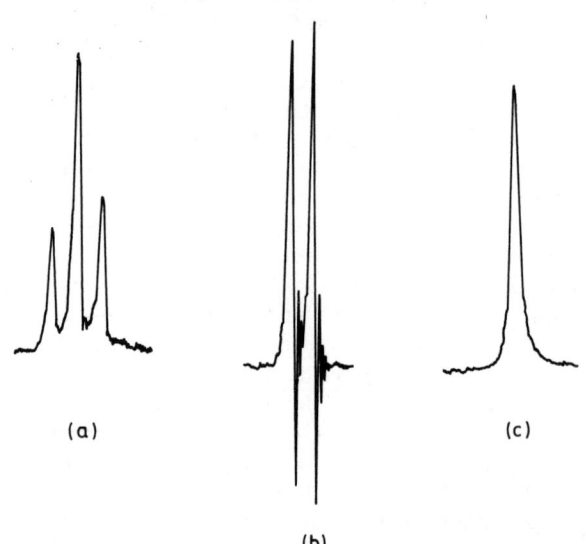

(a)

(b)

(c)

Figure 1.2. The ^{1}H n.m.r. peaks due to —OH groups in alcohols. Spectra taken under conditions of slow exchange of the —OH protons: (a) primary alcohol, (b) secondary alcohol, (c) tertiary alcohol

of nuclear magnetism, and are of interest to the physicist and physical chemist. Texts by Pople, Schneider, and Bernstein[3] and by Emsley, Feeney, and Sutcliffe[4] deal in detail with high resolution n.m.r. spectroscopy, covering background theory, analysis of spectra, and surveying major applications in chemistry. Rather less concerned with mathematical detail, but providing a wealth of numerical data about chemical shifts and coupling constants, the book by Jackman and Sternhell[5] is an enlarged and revised version of the work originally written by Jackman[6]. An introduction to analysis of spectra has been given by Roberts[7], while an advanced account of the mathematical techniques and quantum mechanical ideas required to interpret high resolution n.m.r. spectra is contained in the book by Corio[8]. A number of introductory texts have been published, and these are listed in chronological order[9-15]. Two books[16,17] have dealt particularly with practical aspects of the subject. More specialised reviews of topics in n.m.r. spectroscopy are to be found in two serial publications[18,19], while a regular survey of n.m.r. literature appears in *Annual Reports of N.M.R. Spectroscopy*[20]. A number of compilations of n.m.r. spectra have been published, and some of these are included in the final references[21-25].

References

1. ABRAGAM, A., *The Principles of Nuclear Magnetism*, Oxford (1961)
2. ANDREW, E. R., *Nuclear Magnetic Resonance*, Cambridge (1955)
3. POPLE, J. A., SCHNEIDER, W. G., and BERNSTEIN, H. J., *High Resolution Nuclear Magnetic Resonance*, McGraw-Hill (1959)
4. EMSLEY, J. W., FEENEY, J., and SUTCLIFFE, L. H., *High Resolution Nuclear Magnetic Resonance Spectroscopy*, Pergamon (1965)
5. JACKMAN, L. M., and STERNHELL, S., *Applications of Nuclear Magnetic Resonance Spectroscopy in Organic Chemistry*, Pergamon (1969)
6. JACKMAN, L. M., *Applications of Nuclear Magnetic Resonance Spectroscopy in Organic Chemistry*, Pergamon (1959)
7. ROBERTS, J. D., *An Introduction to the Analysis of Spin-spin Splitting in High Resolution Nuclear Magnetic Resonance Spectra*, Benjamin (1961)
8. CORIO, P. L., *Structure of High Resolution Nuclear Magnetic Resonance Spectra*, Academic Press (1966)
9. ROBERTS, J. D., *Nuclear Magnetic Resonance. Applications to Problems in Organic Chemistry*, McGraw-Hill (1959)
10. BIBLE, R. H., *Interpretation of N.m.r. Spectra*, Plenum Press (1965)
11. MATHIESON, D. W. (Ed.), *Nuclear Magnetic Resonance for Organic Chemists*, Academic Press (1967)
12. CARRINGTON, A., and MCLACHLAN, A. D., *Introduction to Magnetic Resonance*, Harper (1967)
13. BECKER, E. D., *High Resolution N.m.r.*, Academic Press (1969)
14. BOVEY, F. A., *Nuclear Magnetic Resonance Spectroscopy*, Academic Press (1969)
15. LYNDEN-BELL, R. M., and HARRIS, R. K., *Nuclear Magnetic Resonance Spectroscopy*, Nelson (1969)
16. CHAPMAN, D., and MAGNUS, P. D., *Introduction to Practical High Resolution Nuclear Magnetic Resonance Spectroscopy*, Academic Press (1966)
17. STAFF OF VARIAN ASSOCIATES, *N.m.r. and E.p.r. Spectroscopy*, Pergamon (1960)
18. EMSLEY, J. W., FEENEY, J., and SUTCLIFFE, L. H. (Eds.), *Progress in Nuclear Magnetic Resonance Spectroscopy*, Pergamon Press (1966)
19. WAUGH, J. S. (Ed.), *Advances in Magnetic Resonance*, Academic Press (1965)
20. MOONEY, E. F. (Ed.), *Annual Reports of Nuclear Magnetic Resonance Spectroscopy*, Academic Press
21. VARIAN ASSOCIATES, *High Resolution N.m.r. Spectra*, 1 (1962), 2 (1963)

22. HOWELL, M. G. (Ed.), KENDE, A. S., and WEBB, J. S., *Formula Index to N.m.r. Literature Data*, Plenum Press Data Division, **1**, references prior to 1961 (1965), **2**, references 1961–62 (1966)
23. BRUGEL, W., *Nuclear Magnetic Resonance Spectra and Chemical Structure*, Academic Press, **1** (1967)
24. BOVEY, F. A., *N.m.r. Data Tables for Organic Compounds*, Interscience, **1** (1967)
25. SZYMANSKI, H. A., and YELIN, R. E., *N.m.r. Band Handbook*, IFI/Plenum (1968)

2 Basic theory of n.m.r. spectroscopy

2.1 Nuclei in a magnetic field

The nuclei of many isotopic species possess an inherent angular momentum or *spin* which can be described in terms of the nuclear spin quantum number I.

For nuclei with spin angular momentum, I can have the values $\frac{1}{2}$, 1, $\frac{3}{2}$ ....; $I = 0$ is also possible, in which case the angular momentum of the nucleus will be zero. Associated with a spinning nucleus is a magnetic dipole which can interact with an external magnetic field. The energy of this interaction is quantised and can be *described* in terms of the component

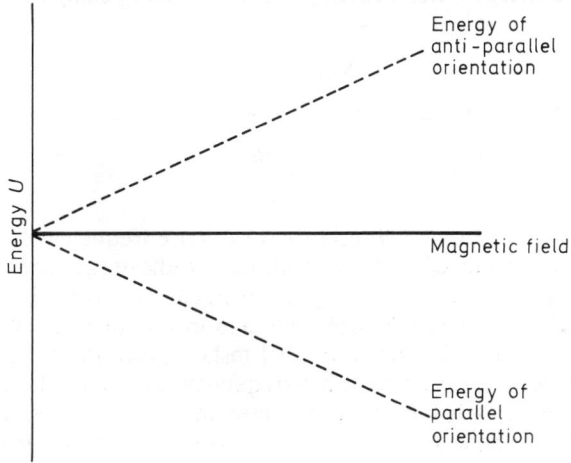

Figure 2.1. The relationship between energy and field strength for the allowed states of a nucleus with $I = \frac{1}{2}$ in a magnetic field

of angular momentum that is parallel to the magnetic field. In general there are $(2I+1)$ possible states, so for the proton $(I = \frac{1}{2})$ there are two possibilities, which can be regarded as corresponding to parallel (lower energy) and anti-parallel (higher energy) orientations of the dipole with respect to the magnetic field. The difference in energy between the two states is linearly related to the field strength as shown in Figure 2.1. If I is greater than $\frac{1}{2}$ there will be more energy states, and in addition the nucleus will have a quadrupole moment arising from an unsymmetrical distribution of nuclear electrical charge. Initially, we shall consider only nuclei for which $I = \frac{1}{2}$ (e.g. ^{1}H, ^{19}F, ^{31}P) paying particular attention to the proton ^{1}H.

For a given magnetic field strength the energy difference between the two states of Figure 2.1 depends on the nuclear species and can be obtained from the relation

$$\Delta U = \frac{\mu B}{I} \tag{2.1}$$

where, when ΔU is the energy difference in joules,

$\qquad\qquad I$ is the nuclear spin quantum number

$\qquad\qquad \mu$ is the magnetic moment of the nucleus in ampere metre^{-1}

and $\qquad\qquad B$ is the magnetic flux density, or the magnetic field, in tesla

I and μ will be constant for a particular nuclear species and are often combined in γ, the magnetogyric ratio defined by

$$\gamma = \mu/I\hbar \tag{2.2}$$

where $\hbar$ is Planck constant divided by 2π.

The units of γ are radians tesla^{-1} second^{-1}. It turns out that it is convenient to measure the energy differences in frequency units ν_0 using the well-known relation

$$h\nu_0 = \Delta U \tag{2.3}$$

so by combining equations (2.1) to (2.3) we obtain the resonance equation

$$\nu_0 = \frac{\gamma B}{2\pi} \tag{2.4}$$

For protons in a field of 1.409 tesla the resonance frequency ν_0 is 60 MHz, which can be produced by conventional radio-frequency techniques. Resonance frequencies for other important nuclei at this field strength are: ^{13}C 15.09; ^{19}F 56.45; ^{31}P 24.29 MHz; and a more complete list of frequencies for field strengths of both 1.409 and 2.301 tesla is given in Appendix 1.

If the nucleus is exposed to radio-frequency energy of the appropriate frequency, transitions will be induced between the two states, and because the population of the lower state is *slightly* (approximately one part in 10^5) greater than that of the upper, there will be a net absorption of energy. A nuclear magnetic resonance spectrometer thus consists of a means of producing a strong magnetic field, a source of radio-frequency power, and a means of detecting absorption of energy by the sample.

2.2 Relaxation

Since the *net* absorption of energy depends on the existence of a finite population difference between the lower and upper states, and since the absorption process tends to diminish this difference, it might be expected that the absorption of energy would soon cease. In fact the upper state is able to lose energy without emitting radiation, and the ways in which it does this are known as relaxation processes. It is necessary to consider these briefly in order to appreciate saturation (p. 55) and certain other phenomena.

There are two relaxation mechanisms available to the nucleus and, in the first of these, energy is lost directly to the surroundings which are referred to as the lattice (even in the case of liquid or gaseous specimens). This process can be described by a rate constant, and T_1 is known as the spin-lattice, or longitudinal, relaxation time. For nuclei with $I = \frac{1}{2}$ in liquids, values of T_1 range from 10^{-2} to 10^2 s unless paramagnetic species are present, in which case T_1 may be much smaller. The second relaxation mechanism is one in which energy is transferred between spins in the system, and is characterised by another rate constant: T_2 is the spin-spin, or transverse, relaxation time. The terms longitudinal and transverse arise when the phenomenological equations derived by Bloch[1] are used to discuss the resonance problem. For most workers the importance of the relaxation times T_1 and T_2 is that they set a limit to the rate at which radio-frequency energy can be supplied to the sample, and hence the strength of signal that can be recorded.

2.3 High resolution n.m.r. spectra

If all the nuclei of a particular species absorbed energy at exactly the same frequency in a given magnetic field, n.m.r. spectroscopy would be of little interest to chemists. However, there are two phenomena—the chemical shift and spin-spin coupling—which can cause absorption to occur over a range of frequencies, and so give valuable structural information. A third phenomenon, exchange, can modify the appearance of n.m.r. spectra; these changes in line *shapes* can be interpreted to give values of kinetic parameters.

The chemical shift

In general the magnetic field (B of equations (2.1) and (2.4)) experienced by a particular nucleus differs from the field of the laboratory magnet. This is because of:

    (a) local fields due to the magnetic dipoles of neighbouring molecules. These can be averaged to zero by the use of isotropic samples (usually liquids) and this is normally done in high resolution n.m.r. spectroscopy

    (b) the magnetic susceptibility of the sample itself. This will affect all nuclei equally and will be discussed later

    (c) local magnetic fields in the same molecule as the nucleus.

In a diamagnetic molecule this last effect is caused by the electronic circulation induced by the main field B_0, and so is proportional in strength to B_0. By Lenz's law these fields will be in such a direction as to oppose the main field, and they effectively screen or shield the nucleus from the complete effect of the applied field.

We may write:

$$B = B_0(1-\sigma) \qquad (2.5)$$

where B is the actual polarising field experienced by the nucleus, and σ is a dimensionless constant called the shielding constant for the nucleus in a particular environment. σ is independent of polarising field strength but

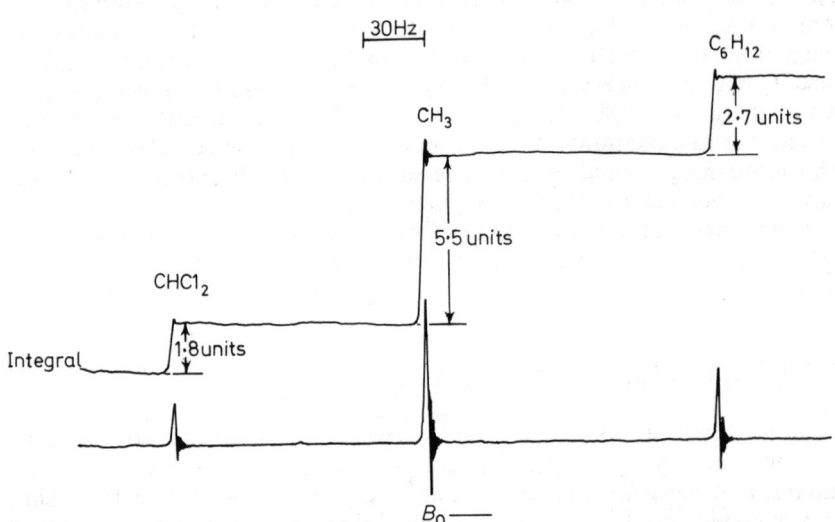

Figure 2.2. Hydrogen n.m.r. spectrum and integral* for a mixture of methyl dichloroacetate and cyclohexane. The relative number of moles of constituents can be obtained from the relative heights of the steps in the integral trace. Comparison of the C_6H_{12} step with that for the $CHCl_2$ proton gives

$$\frac{\text{number of moles } C_6H_{12}}{\text{number of moles } CHCl_2COOCH_3} = \frac{2.7}{1.8} \times \frac{1}{12} = \frac{1}{8}$$

Similarly, comparison of the C_6H_{12} step with that for the CH_3 protons gives

$$\frac{\text{number of moles } C_6H_{12}}{\text{number of moles } CHCl_2COOCH_3} = \frac{2.7}{5.5} \times \frac{3}{12} = \frac{1}{8}.$$

The mole fraction of cyclohexane in the mixture is therefore 0.11

does depend on the chemical environment of the nucleus. It is usually quoted in parts per million (p.p.m.). For nuclei having $I = \frac{1}{2}$, equation (2.4) can now be written as:

$$\nu_0 = \gamma B_0(1-\sigma)/2\pi \qquad (2.6)$$

so nuclei of one species in different environments (i.e. having different values of σ) will resonate at different frequencies if the magnetic field strength B_0 is kept constant.

Alternatively, if the frequency is fixed, different magnetic field strengths will be required to achieve resonance for nuclei in different environments.

* See Sections 4.3 and 5.5.

A spectrum can thus be scanned either by keeping v_0 constant and sweeping B_0 or vice versa. It is customary to *describe* spectra as though they were obtained by varying B_0, and a nucleus for which σ is large will require a larger value of B_0 to bring it into resonance. Thus well-shielded nuclei (i.e. large σ) are said to resonate at high field. As a consequence of the method used to calibrate spectra (Chapter 4) separations between peaks are initially measured in Hz. When, as a result of shielding effects, a number of peaks appear in the n.m.r. spectrum of a particular isotope, the separation between any pair of peaks (ΔB or Δv) depends on the field strength (or the frequency) used in the experiment, so that doubling the applied field will double the value of ΔB or Δv. For convenience in comparing data obtained from instruments working at different field strengths and frequencies, a field independent quantity, δ, is used to express these separations or chemical shifts. If the separation between peaks is ΔB tesla or Δv MHz, when the applied field is B_0 tesla and the frequency is v_0 MHz, then the chemical shift, δ, in p.p.m. between the peaks is

$$|\delta| = \frac{|\Delta B|}{B_0} \times 10^6 = \frac{|\Delta v|}{v_0} \times 10^6$$

This topic will be discussed further in Chapter 4, in the section on referencing.

The signal strength is proportional to the number of nuclei so the spectrum gives information about the relative numbers of nuclei in different chemical environments. Figure 2.2 shows the proton spectrum of a mixture of methyl dichloracetate and cyclohexane. The $3:1$ intensity ratio for the CH_3 and $CHCl_2$ resonances can be seen clearly, and the mole fraction of cyclohexane can be estimated to be 0.11.

The details of the theory of the origin of chemical shifts will not be considered here; however, many empirical correlations are known and some of the more important are listed elsewhere. Typical ranges for chemical shifts are: ^1H in organic compounds *ca* 12 p.p.m.; ^{13}C 400 p.p.m.; ^{14}F 400 p.p.m.; ^{31}P 700 p.p.m.; ^{119}Sn 2000 p.p.m.; ^{207}Pb 20 000 p.p.m.

Spin-spin coupling

The high resolution spectra of many compounds display fine structure which is due to interaction (via the extranuclear electrons) between the spins

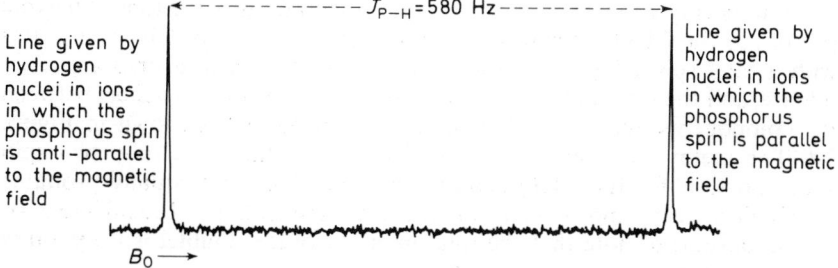

Line given by hydrogen nuclei in ions in which the phosphorus spin is anti-parallel to the magnetic field

$J_{P-H} = 580\ Hz$

Line given by hydrogen nuclei in ions in which the phosphorus spin is parallel to the magnetic field

$B_0 \longrightarrow$

Figure 2.3. ^1H n.m.r. spectrum of the phosphite anion $HPO_3^{-\,-}$

of nuclei in the same molecule. In the phosphite anion HPO_3^- the proton resonance is a doublet, with a separation between the two components of 580 Hz as shown in Figure 2.3, and the ^{31}P spectrum similarly consists of two lines with the same separation. This feature arises because the energy gap between the two spin states of the proton in HPO_3^{--} depends on the spin orientation of the phosphorus nucleus. One of the proton lines (actually the one at higher field) is given by ions in which the phosphorus spin is aligned parallel to the magnetic field, the other comes from ions in which the phosphorus spin is opposed to the field. The orientation of the proton spin affects the phosphorus spin energy gap in the same way. This phenomenon is known as spin-spin splitting and its magnitude—obtained from the separation between the two lines of either doublet—is called the coupling constant.

Coupling constants are measured in Hz (strictly they have the dimensions of energy) and are given the symbol J; they are independent of the applied magnetic field strength.

The spectrum of acetaldehyde is an example of a more complicated situation because the orientations of the individual methyl protons can be combined in several ways. These are given in Table 2.1.

Table 2.1 SPIN STATES OF THE PROTONS IN THE METHYL GROUP OF ACETALDEHYDE

Notations describing spin arrangements	Total spin	Fraction of molecules with particular total spin
$\alpha\alpha\alpha$ or ↑↑↑ or $+++$	$+\frac{3}{2}$	$\frac{1}{8}$
$\alpha\alpha\beta$ or ↑↑↓ or $++-$ $\alpha\beta\alpha$ or ↑↓↑ or $+-+$ $\beta\alpha\alpha$ or ↓↑↑ or $-++$	$+\frac{1}{2}$	$\frac{3}{8}$
$\beta\beta\alpha$ or ↓↓↑ or $--+$ $\beta\alpha\beta$ or ↓↑↓ or $-+-$ $\alpha\beta\beta$ or ↑↓↓ or $+--$	$-\frac{1}{2}$	$\frac{3}{8}$
$\beta\beta\beta$ or ↓↓↓ or $---$	$-\frac{3}{2}$	$\frac{1}{8}$

The probabilities given in the last column of this table are not quite correct because population differences between the upper and lower levels have been ignored; however, the error involved is negligible. Each of the four possible values of the total spin of the methyl protons gives rise to a different energy gap between the two spin orientations of the lone aldehydic proton, so the CHO resonance consists of four lines. Since CH_3 spin states with a total spin of $\pm\frac{1}{2}$ are more probable the two central lines are more intense, and the resonance is actually a quartet with equal spacings between the components and a $1:3:3:1$ intensity distribution as shown in Figure 2.4. The methyl resonance is a doublet because there are only two spin states possible for the CHO proton, and since these are equally probable (to a high degree of approximation) the two components are of equal intensity. It is important to note that the *total* intensity of the doublet is three times that of the quartet.

More complicated cases can be worked out using the rule that in a molecule

AY_n (all nuclei having $I = \frac{1}{2}$) the n equivalent Y nuclei will split the A resonance into $n+1$ lines, whose separation will be equal to the coupling constant J. The relative intensities *within* such a spin multiplet are given by

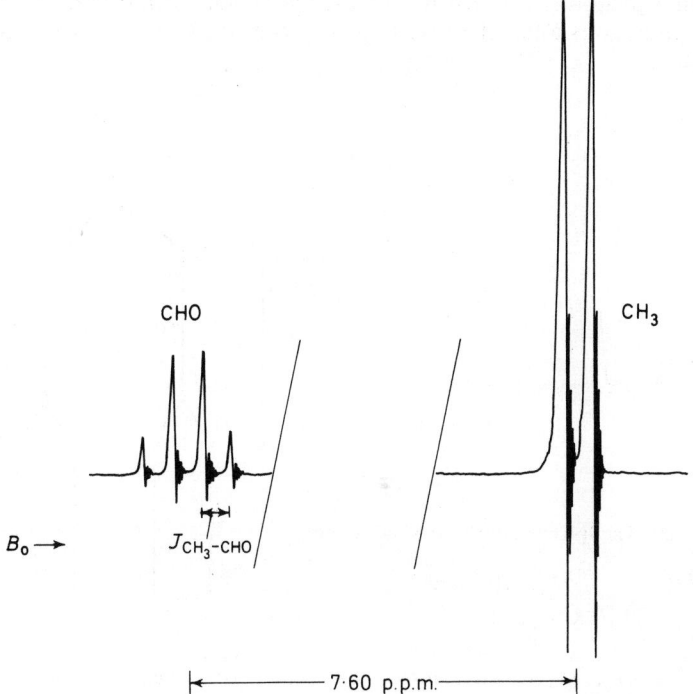

Figure 2.4. The ^{1}H n.m.r. spectrum of acetaldehyde

the coefficients of x in the expansion of $(1+x)^n$, or may be obtained from the appropriate row in Pascal's triangle shown in Table 2.2. If a nucleus A is coupled to more than one group of nuclei (e.g. in a molecule $AY_n Z_m$) the resulting splitting pattern can be obtained by considering first the effect of one group (Y) which will split the A resonance into $n+1$ peaks, then the effect of the second group (Z) will be to split each of the $n+1$ components into a multiplet of $m+1$ peaks.

Table 2.2 COEFFICIENTS IN THE EXPANSION OF $(1+x)^n$, PASCAL'S TRIANGLE

n	Coefficients in expansion									Sum of coefficients for given value of n
0					1					1
1				1		1				2
2			1		2		1			4
3			1	3		3	1			8
4		1		4	6	4		1		16
5		1	5		10	10	5	1		32
6	1		6	15	20	15	6		1	64
7	1	7		21	35	35	21	7	1	128
8	1	8	28	56	70	56	28	8	1	256

For example the resonance of the α proton in vinyl acetate is a doublet of doublets as a result of splittings of 6.5 and 14.1 Hz by the protons *cis* and *trans* to it as shown in Figure 2.5 (a). Figure 2.5 (b) shows a second example in which a proton is coupled to two methyl groups with different values of the coupling constants; the resulting spectrum for this proton is a quartet of quartets.

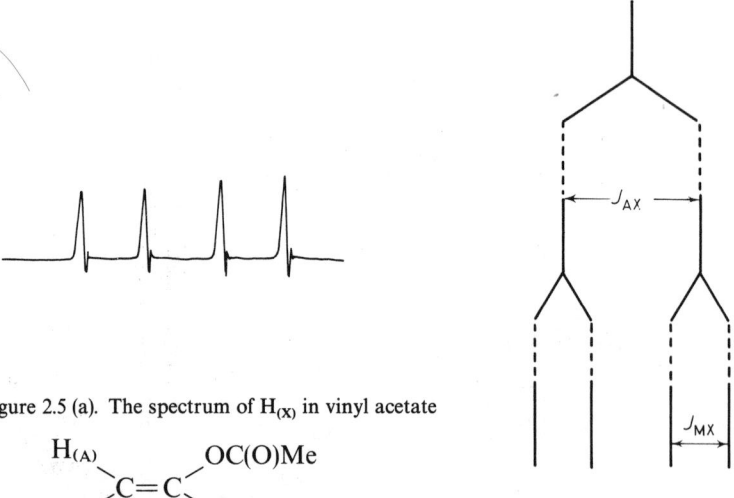

Figure 2.5 (a). The spectrum of $H_{(X)}$ in vinyl acetate

$$\begin{array}{c}H_{(A)} \qquad\qquad OC(O)Me \\ \diagdown C{=}C \diagup \\ \diagup \qquad\qquad \diagdown \\ H_{(M)} \qquad\qquad H_{(X)}\end{array}$$

in which $J_{AX} > J_{MX}$. The line diagram shows the splitting pattern built up by successive consideration of individual couplings

These simple rules for calculating the patterns, which arise from spin-spin splitting, apply only when the chemical shift separation Δv (measured in Hz) is large compared with the magnitudes of the coupling constants. Such spectra are said to be first order, and are most likely to arise when the measuring frequency, v_0, is large. This is one of the main reasons for the popularity of high magnetic field strengths in n.m.r. spectroscopy. In other cases second-order theory must be used, but the discussion of this is beyond the scope of this book. Useful introductions are given in references 2, 3 and 4, and fuller details are in references 5 to 7.

As an example of the kind of behaviour to be expected, Figure 2.6 shows the spectrum given by an ethyl group for different values of the chemical shift separation between the methyl and methylene resonances. In Figure 2.6 (a) an essentially first-order triplet and quartet are obtained, but as Δv decreases the intensity pattern becomes unsymmetrical as in (b). In Figure 2.6 (c) additional lines appear, and in (d) it is no longer easy to assign the lines as either methyl or methylene resonances. For many purposes it is probably permissible to extract δ and J from spectra (a) to (c) by means of first-order analysis, and it is useful to note that as δ decreases the intensities increase on the side of the multiplet nearest the resonance of the coupled group.

A convenient scheme of spectral classification has been introduced to describe more complicated systems, and is as follows. Capital letters are used to denote nuclei within a coupled spin system; a different letter is used for chemically non-equivalent nuclei. If the relevant nuclei are tightly coupled, i.e. if the chemical shifts and coupling constants are of comparable

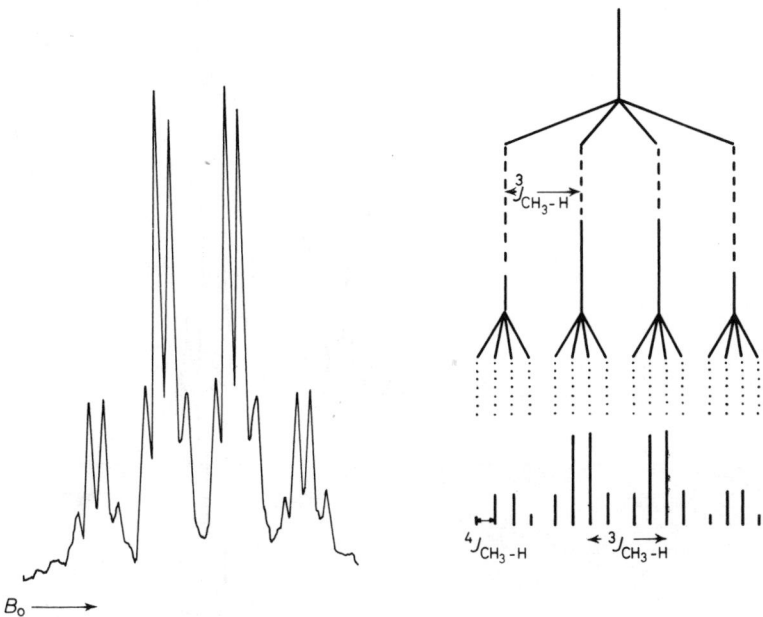

B_0 ⟶

Figure 2.5 (b). The spectrum of $H_{(X)}$ in α,β-dimethylacrolein,

$$CH_3CH_{(X)} = CCH_3(CHO)$$

in which $^3J_{CH_3 - H_{(X)}} > {}^4J_{CH_3 - H_{(X)}}$. The line diagram shows the splitting pattern built up by considering the couplings successively.

In the symbol $^3J_{CH_3 - H_{(X)}}$ the superscript indicates that there are three bonds between the coupled nuclei, i.e.

$$H\!-\!\overset{\displaystyle |}{C}\!-\!\underset{H_{(X)}}{\overset{\displaystyle |}{C}}\!=$$

while $^4J_{CH_3 - H_{(X)}}$ refers to the coupling

$$\underset{H_{(X)}}{\diagup}C = C\!-\!C\!-\!H$$

in which there are four bonds between the coupled nuclei

magnitude, then letters close together in the alphabet are used, whilst if the coupling is weak (i.e. the coupling constants are much smaller than the chemical shifts measured in Hz) then letters far apart in the alphabet are used. Thus in $CH_3CH_2SiCl_3$ there is a small chemical shift difference between the methyl and methylene protons, and the system is A_3B_2, but in CH_3CHO there is a large chemical shift difference so the description A_3X

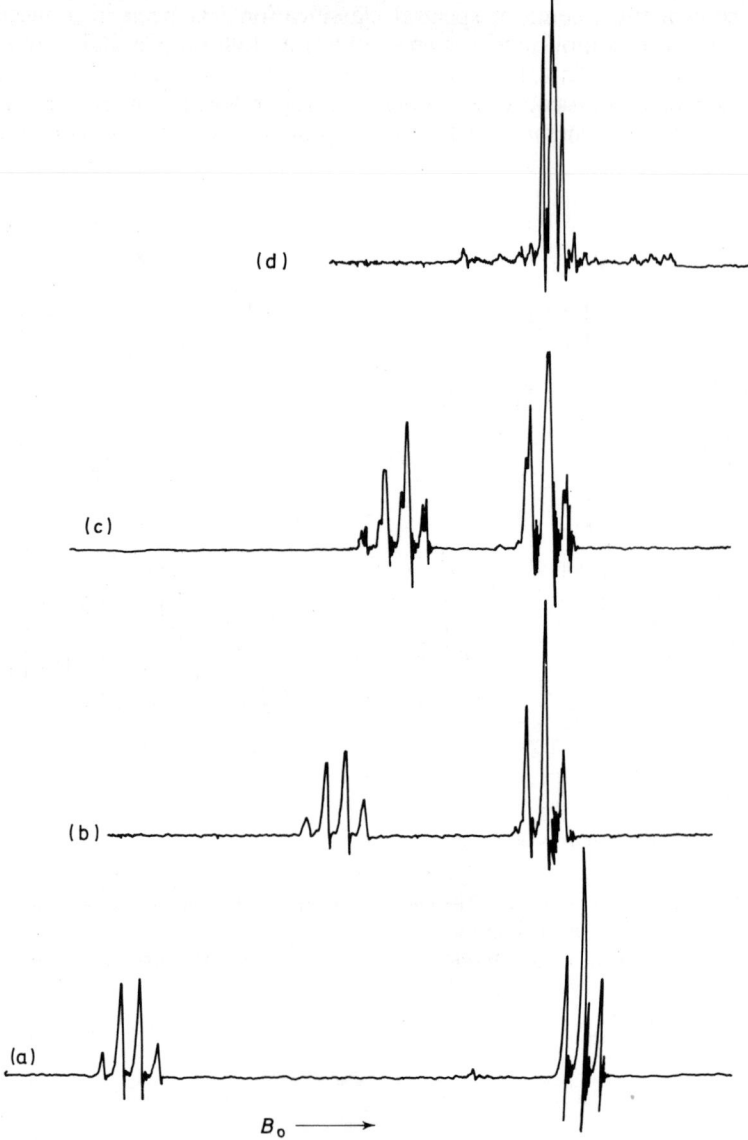

Figure 2.6. Hydrogen n.m.r. spectra of ethyl compounds.

(a) $CF_3COOCH_2CH_2$ ($\Delta v/J \sim 25$). This spectrum is essentially first order, and consists of a $1:3:3:1$ quartet together with a $1:2:1$ triplet.

(b) $C_6H_5CH_2CH_3$ ($\Delta v/J \sim 12$). With decreasing chemical shift the intensity pattern becomes unsymmetrical, peaks towards the centre of the spectrum gaining intensity at the expense of peaks to the outside of the spectrum.

(c) $Te(CH_2CH_3)_2$ ($\Delta v/J \sim 8$). As the chemical shift is further decreased additional lines appear: the intensity pattern becomes increasingly unsymmetrical.

(d) $Sn(CH_2CH_3)_2(C\equiv CH)_2$ ($\Delta v/J \sim 1$). It is no longer easy to assign the lines as either methyl or methylene resonances, or to detect the quartet and triplet features found in the examples (a) to (c) may be used. A nucleus of a different species always introduces some genuinely first-order features, so $CH_3CH_2PCl_2$ gives an A_3B_2X spectrum, and CH_3PFBr is an A_3MX system.

If there is no coupling between certain nuclei in different parts of a molecule it may be justifiable to use a simpler description than is strictly correct. For example, the proton spectrum of triethyl phosphine, $(CH_3CH_2)_3P$, has been analysed satisfactorily as part of an A_3B_2X system, because there is essentially no interaction between different ethyl groups in the same molecule.

Difficulties may arise when certain nuclei are chemically but not magnetically equivalent. The tetrahedral molecule CH_2F_2 is correctly described as an example of an A_2X_2 spin system, because there is only one type of H–F coupling constant. However in 1,2-difluoro-4,5-dichlorobenzene

$$\begin{array}{c} Cl \quad Cl \\ H\langle \bigcirc \rangle H \\ F \quad F \end{array}$$

there are two different H–F couplings (i.e. the one between an H and an F that are mutually *ortho*, and the one where the two nuclei are mutually *meta*) and the system must be described as AA'XX', where the dashes serve to distinguish nuclei that have identical chemical shifts but are coupled differently to other nuclei in the spin system. In these circumstances the observed spectrum depends on the four coupling constants $J(AX)$, $J(AX')$, $J(AA')$, and $J(XX')$, whereas in the true A_2X_2 case the A–A and X–X couplings have no effect at all on the appearance of the spectrum.

In recent papers[8, 9] the notations for n.m.r. spin systems have been summarised, their disadvantages have been indicated and a new notation has been proposed.

Exchange processes

So far it has been assumed that the chemical environment of the nucleus remains unchanged during the n.m.r. experiment. If this is not so the spectrum obtained will depend upon the time the nucleus spends in the different environments. A simple case is one in which protons exchange between two sites which are occupied for equal lengths of time. If the lifetime in the two sites is long, then two separate signals will be obtained, each with a chemical shift that corresponds to the shielding of the proton at the appropriate site. As the rate of exchange is increased (say by adding a catalyst or raising the temperature) so that the lifetimes become shorter, it is found that the two signals broaden, approach one another, then coalesce to a single broad line, and finally give a sharp singlet when the exchange rate is rapid enough. The chemical shift of the singlet is the mean of those of the two lines obtained in the slow-exchange situation, and the overall behaviour is illustrated in Figure 2.7.

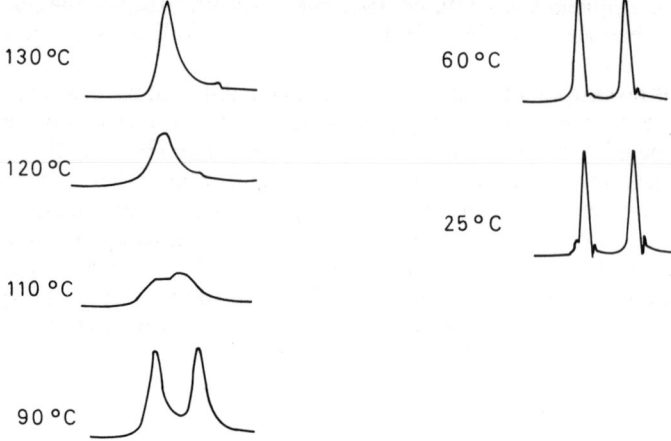

Figure 2.7. Changes in ^{1}H n.m.r. spectrum as the rate of proton exchange between two sites is increased by increasing the temperature. The two sites are occupied for equal times. If the temperature of the sample were raised above 130°C the singlet would become sharper

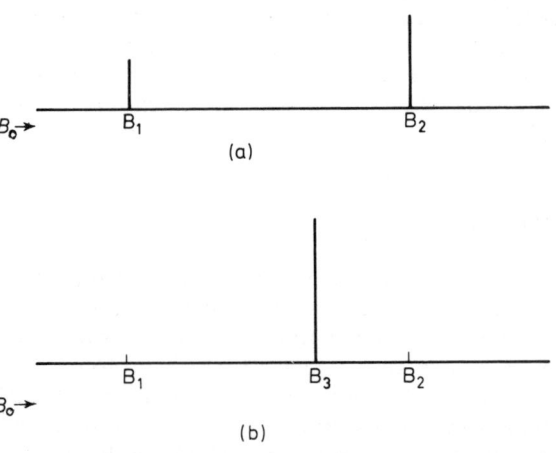

Figure 2.8. Diagrammatic representation of the change in spectrum which occurs on increasing the rate of exchange of protons between two sites when the populations of the two sites are different: (a) slow exchange, (b) rapid exchange. The figures represent an idealised form for the exchange between H_2O and the hydroxylic proton in a tertiary alcohol ROH. The populations in the two states being $2:1$ for a mixture containing equal numbers of moles of H_2O and ROH. It is then found that $(B_3 - B_1):(B_2 - B_3) = 2:1$

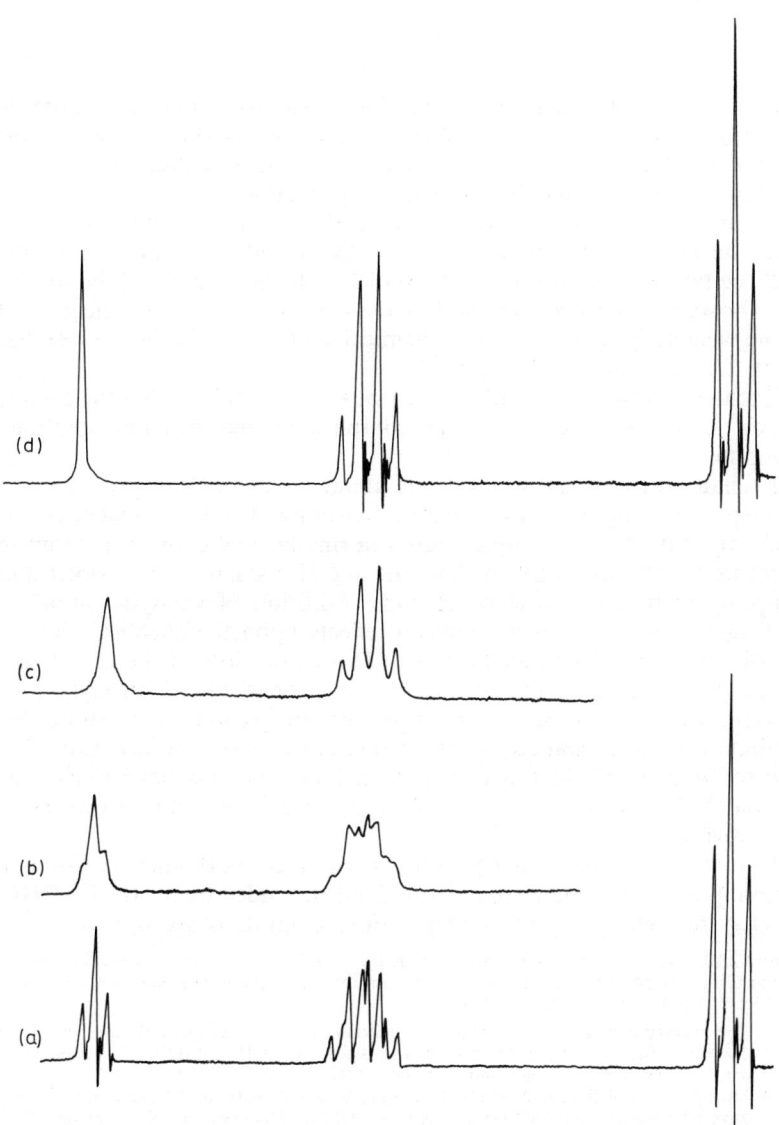

Figure 2.9. Effect of increasing the rate of exchange of the —OH proton in ethyl alcohol. In (a) the rate of exchange is slow. By addition of catalytic amounts of acid the rate of exchange is increased, (b) and (c) showing broad patterns for the OH and CH₂ peaks. (d) shows the sharp peaks obtained when the rate of exchange becomes very rapid

The exchange rate when the central minimum between the two separate signals just vanishes (this is sometimes known as the coalescence point) is given by Eq. (2.7)

$$\tau\Delta v = \frac{1}{\pi\sqrt{2}} \tag{2.7}$$

where Δv is the chemical shift separation of the two resonances in the slow-exchange limit, and 2τ is the residence time at *either* site. It is important to appreciate that the normal rate constant for the exchange process is then $\frac{1}{2}\tau$. Owing to the difficulty of estimating coalescence points exactly, Eq. (2.7) can only be regarded as a guide to the rate of exchange, and a more accurate value may be found from a detailed consideration of the line-shapes.

If the two sites are not equally populated the behaviour will be similar to the above, but the two lines will coalesce to a point whose chemical shift is the weighted mean of the two chemical shifts found in the slow-exchange limit. This is illustrated in Figure 2.8.

Exchange between several sites may be treated similarly, but the equations which arise are generally rather complicated and demand solution by computer.

Exchange processes can also bring about the loss of fine structure due to spin-spin coupling. The hydroxyl proton in ethanol undergoes facile chemical exchange, but in pure samples this is normally slow enough at room temperature for the spectrum to show the —OH resonance as a triplet due to coupling with the methylene protons. Addition of catalytic amounts of hydrogen ions produces insignificant effects upon the chemical shifts, but greatly accelerates the exchange rate. This causes loss of the fine structure as shown in Figure 2.9 (d), because the time for which a hydroxylic proton is associated with a *particular* ethoxy group becomes very short. Intermediate rates of exchange give the broad patterns shown in Figures 2.9 (b) and (c). A detailed treatment of this exchange process has been given by Arnold[10]. The uses of n.m.r. in the study of chemical rate processes have been reviewed by Johnson[11].

In many cases the exchange process will affect both chemical shifts and coupling constants and a simple example is provided by the $(CH_3)_2CHOH/H_2O$ system whose spectra under various conditions are shown in Figure

Figure 2.10. Exchange of —OH protons in the *iso*-propyl alcohol/water system. The peaks due to the CH_3 groups of the alcohol have not been included; these give rise to a sharp, intense doublet which does not change in appearance.

(a) The spectrum of the —OH and CH protons of *iso*-propyl alcohol. The sharp —OH doublet indicates that the rate of exchange is slow; the CH multiplet arises from coupling with the two methyl groups and with the —OH proton.

(b) The spectrum of the same sample of alcohol to which water and a small amount of acid have been added. Separate peaks due to —OH and H_2O protons are observed; the rate of exchange is sufficient to cause some broadening of the peaks.

(c) The rate of exchange has been increased by addition of a further small quantity of acid. The effect of coupling between —OH and CH groups has just about been lost, the coupling between the CH_3 and CH groups is still detectable.

(d) On further increasing the rate of exchange the —OH and H_2O peaks begin to broaden and merge. The CH multiplet, showing the splitting due to coupling with the CH_3 groups is becoming sharper as the frequency of exchange becomes much greater than J_{CH-OH}.

(e) The rate of exchange is sufficiently rapid for the —OH and H_2O protons to be indistinguishable. A single peak appears at a weighted mean position for these protons. The CH multiplet has sharpened further

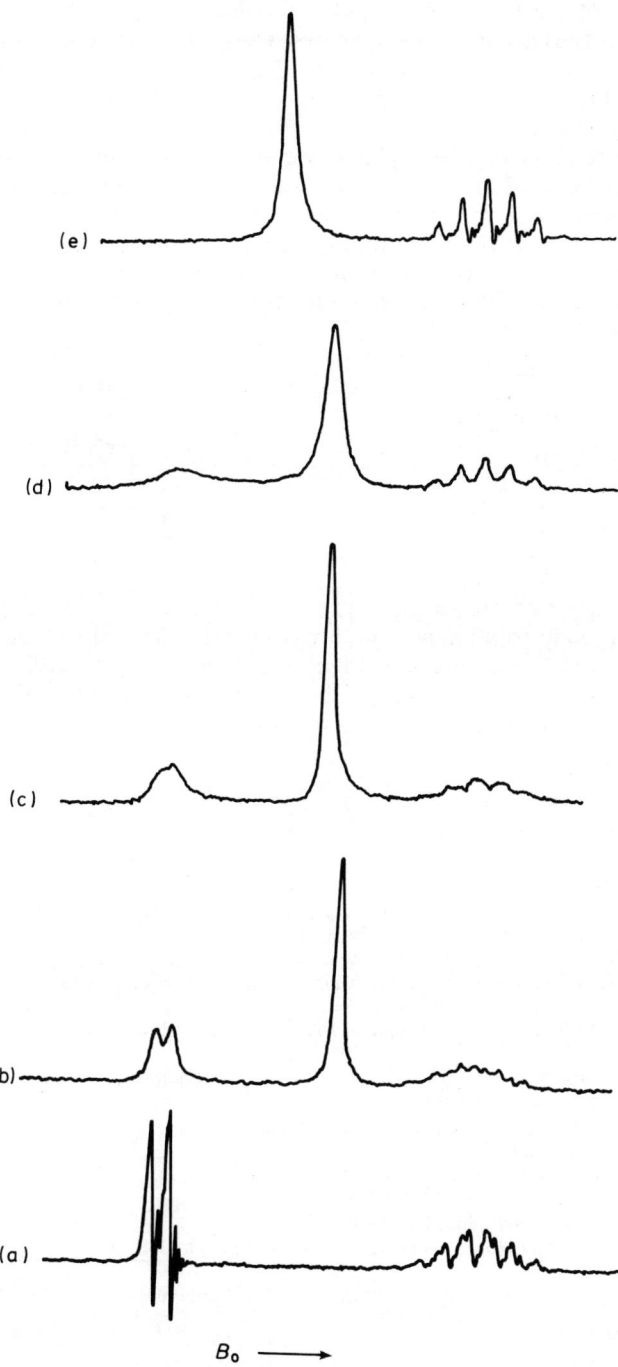

$B_0 \longrightarrow$

2.10. Here can be seen the effects of (a) loss of spin-spin multiplicity owing to exchange of the OH proton, and (b) coalescence of the H_2O and OH resonances as the exchange rate increases. The first occurs at a relatively low exchange rate since it is governed by $J_{CH-OH} = 6$ Hz, while the second demands a higher rate of exchange since it depends on a chemical shift difference of 40 Hz.

Rapid interchange between different molecular conformations can lead to averaging of chemical shifts in a similar way, but since bonds are not broken the fine structure due to spin coupling is generally retained. However, the coupling constants may differ among conformations of a given molecule; the observed coupling constants in the fast inversion situation will be *algebraic* averages of those in the individual conformers. For example, in the molecule Cl_2PF_3

exchange of fluorine atoms between axial (a) and equatorial (e) positions occurs by a pseudo rotation without breaking of bonds. At low temperatures the process is slow and the individual couplings $J_{P-F(a)}$ $(= 1023$ Hz) and $J_{P-F(e)}$ $(= 1085$ Hz) can be distinguished. At room temperature the process is rapid and only an average J_{P-F} can be measured[12].

References

1. BLOCH, F., *Phys. Rev.*, **70**, 460 (1946)
2. ROBERTS, J. D., *An Introduction to the Analysis of Spin-spin Splitting in High Resolution Nuclear Magnetic Resonance Spectra*, Benjamin (1961)
3. BECKER, E. D., *High Resolution N.m.r.*, Academic Press (1969)
4. LYNDEN-BELL, R. M., and HARRIS, R. K., *Nuclear Magnetic Resonance Spectroscopy*, Nelson (1969)
5. CORIO, P. L., *Structure of High Resolution Nuclear Magnetic Resonance Spectra*, Academic Press (1966)
6. POPLE, J. A., SCHNEIDER, W. G., and BERNSTEIN, H. J., *High Resolution Nuclear Magnetic Resonance*, McGraw-Hill (1959)
7. EMSLEY, J. W., FEENEY, J., and SUTCLIFFE, L. H., *High Resolution Nuclear Magnetic Resonance Spectroscopy*, Pergamon (1965)
8. AULT, A., *Journal of Chemical Education*, **47**, 812 (1970)
9. HAIGH, C. W., *J. Chem. Soc.* (A), 1682 (1970)
10. ARNOLD, J. T., *Phys. Rev.*, **102**, 136 (1956)
11. JOHNSON, C. S., *Advances in Magnetic Resonance*, **I**, Edited by J. S. Waugh, Academic Press (1965)
12. MUETTERTIES, E. L., and co-workers, *Inorg. Chem.*, **2**, 613 (1963) and **3**, 1298 (1964)

3 Interpretation of hydrogen n.m.r. spectra

3.1 Introduction

Spectral interpretation may be considered to embrace all aspects of extracting information from an n.m.r. spectrum, that is, it ranges from obtaining the n.m.r. parameters (chemical shifts and coupling constants) to using the technique to establish molecular structure. The former is generally referred to as spectral *analysis* and is easy only for first-order or *quasi* first-order spectra. The latter aspect is generally the more valuable for practising chemists, but depends to a considerable extent on a successful analysis of the spectrum. The basic problem is to distinguish between separations due to spin-spin splitting, and separations due to chemical shifts; these are then assigned to interactions and groups respectively in the proposed structure. Clearly the more information from other sources that is available about the molecule the better, and in a satisfactory interpretation the experimental spectrum will reproduce correctly all the features that the spectrum of a molecule with the proposed structure might be expected to display.

In principle the determination of the structure of a molecule from its n.m.r. spectrum involves a consideration of all possible structures and the spectra they would be expected to give; in practice this procedure can often be simplified. There may be, for example, special features in the observed spectrum which allow immediate elimination of all but one or two alternative structures, even though a complete interpretation might be impossible. It has been our experience that facility in the interpretation of n.m.r. spectra is best acquired by practice, and in this respect the two Varian catalogues[1] are useful as they present a large number of interpreted spectra roughly graded in order of complexity. In Appendix 6 we present detailed interpretations of a number of spectra specially selected to illustrate important points that can arise. These can be used as exercises if desired. The most important weapon in the interpreter's armoury is a knowledge of the chemical shifts and coupling constants likely to arise in particular situations, and the correlation tables in the appendices will be found useful in this respect. Additionally, it is often possible to modify the experimental conditions in such a way that informative changes occur in the appearance of the spectrum, and we now consider some of these techniques.

3.2 Experimental aids to spectral interpretation

Solvent effects[2]

Proton chemical shifts are often quite sensitive to changes of solvent, although coupling constants tend to remain unaffected. Thus if a compound which gives a complicated spectrum is studied in a number of different solvents, changes may occur which aid interpretation; in particular, spectra which are strongly second order in one solvent may become amenable to a first-order analysis in another. In practice it is found that aromatic solvents, such as benzene, are very useful on account of their high diamagnetic anisotropy which can lead to strong differential effects upon the chemical shifts of protons in solute molecules. For example, the proton spectrum of triphenyl phosphine, $(C_6H_5)_3P$, dissolved in deuteriochloroform is a poorly resolved band as

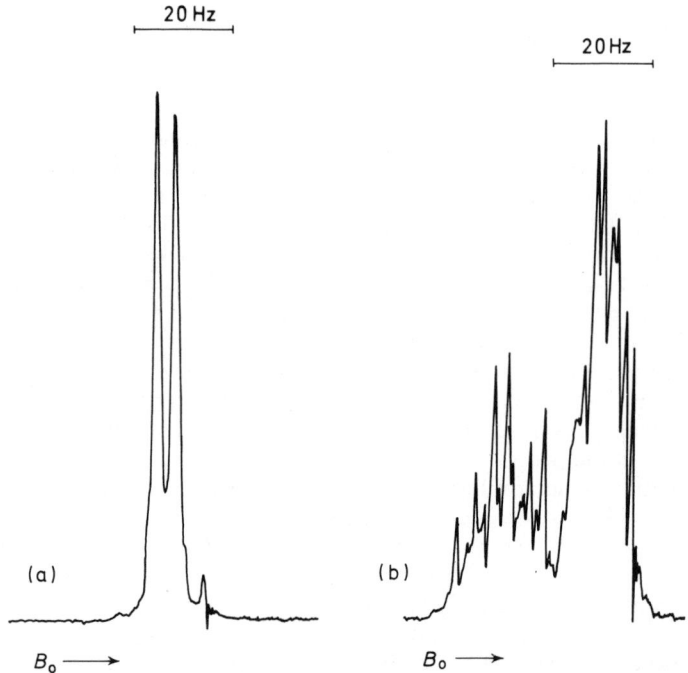

Figure 3.1. The ^1H n.m.r. spectrum of triphenyl phosphine in different solvents, at 60MHz. (a) In deuteriochloroform; no distinction between the aromatic ring protons is possible from this spectrum. (b) In hexadeuteriobenzene; separate regions of absorption occur, the low-field multiplet arising from the protons ortho- to the phosphorus atom

shown in Figure 3.1 (a), but when perdeuteriobenzene is the solvent two separate regions of absorption are found, as in Figure 3.1 (b), and certain coupling constants can be measured.

This ability to induce chemical shifts selectively (the so-called phenomenon of ASIS, Anisotropic Solvent Induced Shifts) is one of the most valuable features of aromatic solvents and can have considerable diagnostic value.

Classic examples are provided by the behaviour of substituted amides in different solvents; thus the methyl resonance of N,N-dimethylformamide

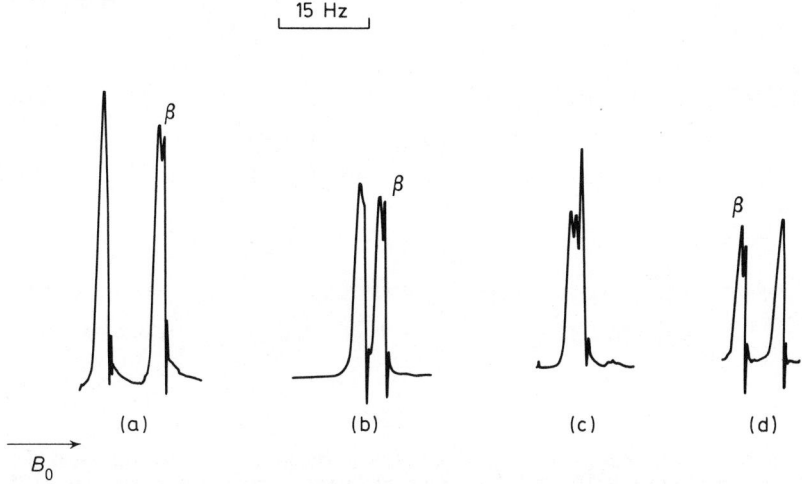

in deuteriochloroform consists of two lines owing to restricted rotation about the N–CO bond which renders the two methyl groups non-equivalent. As benzene is added to the solution (Figure 3.2) the two separate signals approach one another, merge, and finally appear in interchanged positions.

15 Hz

(a)   (b)   (c)   (d)

B_0

Figure 3.2. Anisotropic Solvent Induced Shifts—the effect of benzene on the spectrum of dimethylformamide. The methyl group protons of dimethylformamide in chloroform give the spectrum shown in (a). The high-field peak (β) shows a splitting due to coupling with the CHO proton ($J = 0.65$ Hz). The low-field peak is broadened by coupling to the CHO proton ($J \sim 0.3$ Hz). The difference in coupling serves to distinguish the two peaks in the subsequent spectra. On successive addition of portions of benzene the two peaks draw closer together (b), just exchange relative positions (c), and then separate (with (β) now at lower field) as in (d)

This behaviour is attributed to an association of the type[3] indicated in Figure 3.3 between the amide and benzene molecules, the effect arising

Figure 3.3. Association of an amide with benzene. The +ve end of the molecular dipole of the amide is shown located over the plane of the aromatic ring; the −ve carbonyl group is situated as far away as possible from the centre of the ring while maintaining planar association. In neat amide the protons of the (α) methyl group appear at low field, while in benzene solution these protons are more strongly shielded than those of the (β) methyl group

because the magnetic anisotropy of benzene is such as to increase the shielding of groups above or below the plane of the ring, and to decrease the shielding of groups near the edge of the ring. There is some dispute as to the precise origin of this anisotropy; one popular model is that of an electronic ring current. This model invokes a circulation of the electrons induced by the magnetic field, and it has the advantage that the degree of shielding or deshielding to be expected at any particular point may be calculated easily. The results of such a calculation are presented graphically in Figure 3.4, and there is little doubt about the qualitative validity of this picture whatever its theoretical justification may be.

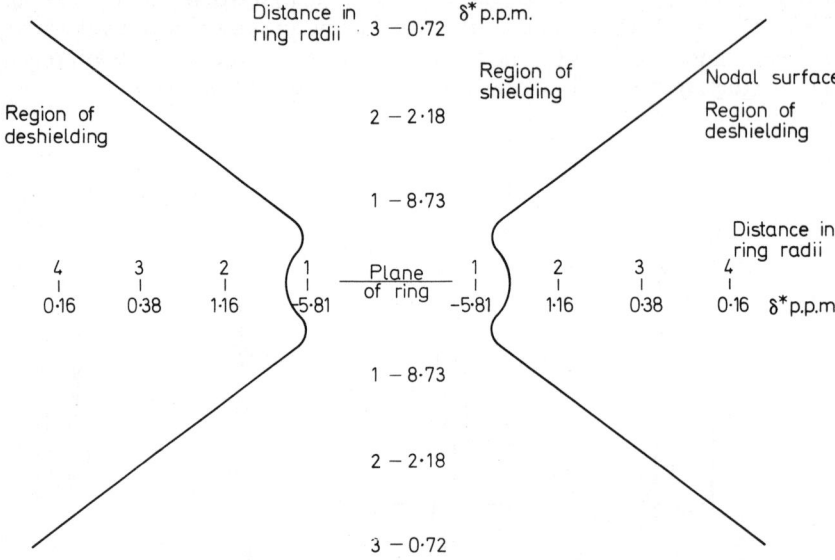

Figure 3.4. The shielding zone about a benzene ring. The heavy line shows the nodal surface between regions of shielding and deshielding. Distances are measured in ring radii (1.39 Å, 0.139 nm), δ^* is the shielding (in p.p.m.) calculated for the ring current arising from the six π electrons. In the calculations allowance has been made for the fact that the π electron cloud has its maximum density in two regions, one on each side of the benzene ring. (Johnson, C.E., and Bovey, F. A., *J. Chem. Phys.*, **29**, 1012 (1958))

The chemical shifts of certain protons in numerous other carbonyl-containing compounds are affected similarly by benzene, and of especial interest are various steroid molecules. The behaviour of these compounds has been studied extensively by Bhacca and Williams, and details are to be found in their book[4]: only one example will be presented here. In the majority of cases the association between the benzene molecule and the carbonyl group is such that protons near the carbonyl groups are out of the plane of the benzene ring, that is they experience increased shielding. However in 5-androstan-11-one the association can be expected to be as shown, and in consequence the C–19 methyl group should be deshielded. This is found to be the case to the extent of 0.14 p.p.m.

The magnetic anisotropy of pyridine is similar to that of benzene, but owing to the presence of the electronegative nitrogen atom, with its lone pair of electrons, this molecule can be expected to adopt a different mode of association with solutes. Thus the pyridine might be oriented perpendicular to the plane of an amide group, and this would lead to deshielding of associated protons as has been observed in certain cases. Of course, in certain circumstances, the use of benzene as a solvent may not have a beneficial effect. An example of this is provided by pyrrole which gives two distinct regions of absorption for the C–H protons in deuteriochloroform solution, but only a single band in benzene.

Another useful effect is the ability of certain solvents to promote rapid exchange of labile hydrogen atoms such as those in hydroxyl and amino groups, thereby simplifying the spectrum given by other protons in the same molecule. Thus in $CDCl_3$ solution the resonance of the methylene protons adjacent to the hydroxyl groups in many straight chain alcohols is a simple triplet; contrariwise, some solvents (e.g. dimethyl sulphoxide) can form strong hydrogen bonds with acidic protons and substantially reduce the rates of

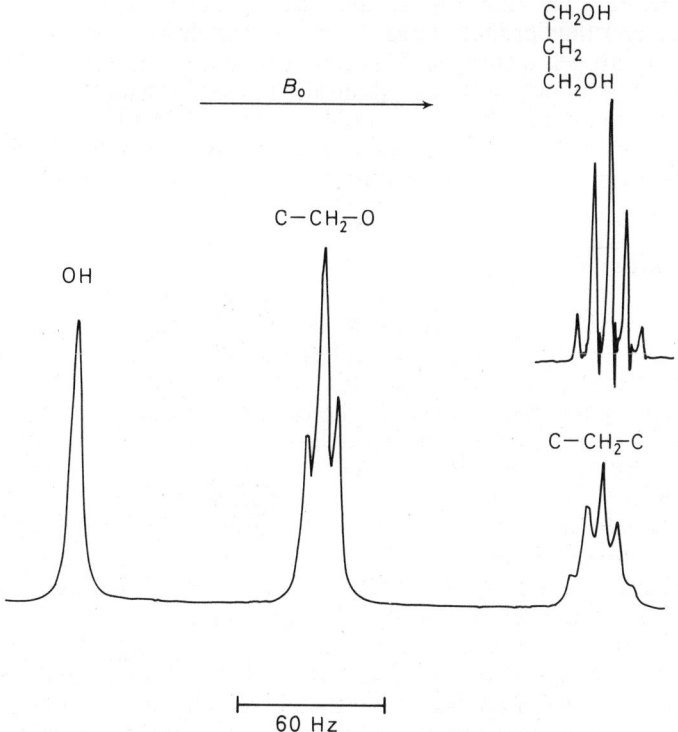

Figure 3.5. Variation of line-width with temperature. The room temperature ^{1}H n.m.r. spectrum of propane-1,3-diol shows broadened lines because of the viscosity of the liquid. The inset shows the improvement in line-width obtained when the viscosity is reduced by heating the sample to 90°C

exchange processes in which they are involved. In this way the hydroxyl protons of even rather wet alcohols can be induced to display spin-spin coupling to other protons in the same molecule.

Temperature variation

Raising the temperature of the sample can often be useful as a means of increasing solubility or reducing the viscosity of a solution (Figure 3.5), but there are also ways in which temperature variation can assist spectral interpretation. Spectra may have broad lines as a result of exchange or conformational inversion processes that are occurring at an intermediate rate (p. 15). If the temperature of such a system is raised the process will be accelerated and in general the lines will ultimately become sharper and their number will decrease, while if the temperature can be reduced sufficiently each individual species that is present may give its own sharp spectrum. An example of this is provided by N,N-dimethyl trichloroacetamide. The coalescence temperature for the methyl peaks in the spectrum of this compound is 287.1 K, so at room temperature the spectrum would be a broad band (cf. N,N-dimethylformamide, Figure 2.7).

Temperature variation can be used also to alter the extent of hydrogen bonding, and thus produce large changes in the chemical shifts of protons bound to nitrogen or oxygen. This can be a convenient way of identifying the resonances of such protons, and can also be used to change the appearance of a spectrum from second to first order. In cases where the solvent contains hydrogen-bonded protons, the choice of one or more suitable temperatures for measurement may avoid obscuring an important part of the spectrum.

Deuteriation

The use of deuteriated solvents to avoid obscuring resonances is discussed later (p. 52), and partial deuteriation of samples can bring about considerable simplification of spectra. Acidic protons (i.e. those attached to oxygen or nitrogen atoms) are the easiest to replace, and this can be done with an extremely high degree of selectivity simply by shaking the sample with D_2O and then drying if necessary. In many cases it may suffice merely to use D_2O as the solvent. Specific deuteriation at a chosen site in the molecule can be used to identify positively particular resonances, and can be specially valuable for following the position of a particular atom in mechanistic studies. It is important to remember that deuterium had a spin quantum number of 1, so the resonances of any protons to which the replaced proton was spin coupled will now be split into $1:1:1$ triplets; however, couplings involving deuterium are $\frac{1}{7}$ of the corresponding proton couplings, because of the smaller magnetogyric ratio of deuterium.

Massive deuteriation can be used to remove the bulk of the protons from a sample, and thus simplify study of the resonances of the remainder, but the cumulative effect of coupling to so many deuterium nuclei may lead to rather broad lines in the proton spectrum. This effect can be eliminated conveniently by deuterium decoupling (p. 75). Substitution of deuterium for hydrogen normally has a small effect upon the chemical shifts of other protons in the molecule and this increases in proportion to the amount of deuteriation. The changes observed are greatest for protons geminal to the deuterium atom and seldom exceed 0.05 p.p.m.

Measurement at a different radio frequency

If the strength of the magnetic field is changed, then coupling constants will remain unaltered while chemical shift differences (as measured in Hz) will change. In the case of a spectrum which is already essentially first order at 60 MHz, a measurement at 100 MHz will thus serve to distinguish unequivocally splittings which arise from spin coupling from those which are due to chemical shift differences. Indeed such a distinction can often be made by a comparison of spectra recorded at say 56.4 and 60 MHz, which may be done on one spectrometer with two different radio-frequency (r.f.) units.

A more important use of magnetic field strength variation is to reduce the degree of second-order character, and it is often found that a spectrum which is intractable at 60 MHz will yield reasonably accurate parameters by first-order analysis at 100 MHz. Spectrometers in which the magnet has a superconducting solenoid, and which operate at a proton resonance frequency of 220 MHz, are now becoming available, and these often give essentially first-order spectra for quite complicated molecules such as steroids.

In a strictly first-order spectrum, information regarding the relative signs of coupling constants is missing, and it can therefore be advantageous to record the spectrum also at a *lower* field strength. Some second-order character may be introduced thereby, and some of the sign information may be recovered.

Spin decoupling

Often the difficulty in interpreting an n.m.r. spectrum in detail arises from an overwhelming mass of fine structure due to numerous spin coupling interactions. The technique of spin decoupling permits the removal of these in a selective manner and so facilitates analysis. The reader is referred to Chapter 5 for the experimental details and limitations of this technique, and Figure 3.6 shows the simplification in the resonance of the olefinic protons of crotonaldehyde that can be accomplished by decoupling the methyl group. This experiment also establishes beyond doubt that the quartet structure displayed by these resonances does indeed arise from coupling to the methyl group. Heteronuclear decoupling experiments can be used similarly, both to simplify proton spectra and to confirm that observed line splitting is due to a particular isotope. This last application of the technique is especially valuable in the case of suspected coupling to nuclei of low natural abundance. In this case the lines of interest appear as weak satellites of the main peak, and these may be difficult to distinguish from impurity lines. A decoupling experiment such as that illustrated in Figure

3.7 for the ^{29}Si satellites of $(Me_3Si)_2O$ can then be used to settle the point.

In the proton spectra of complex molecules resonances of special importance may be completely hidden by signals from other protons in the sample, but if the proton of interest is spin coupled to one whose resonance

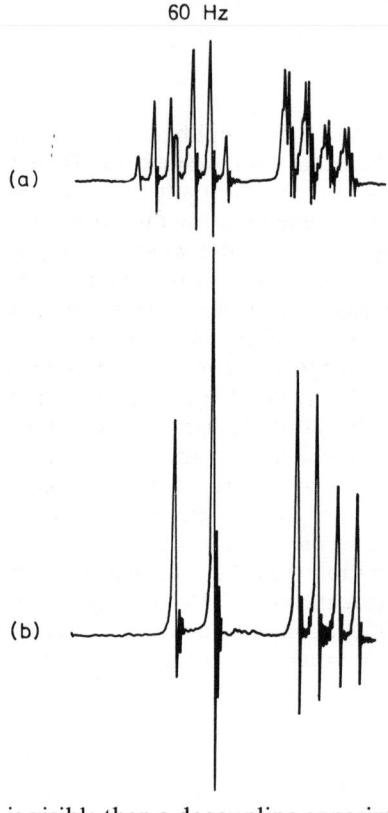

60 Hz

(a)

(b)

Figure 3.6. Simplification of spectra by homonuclear decoupling.

(a) Shows the olefinic proton spectrum of crotonaldehyde:

The low-field multiplet arises from the proton (α) and exhibits coupling with the methyl group (to give a quartet) and with the proton (β) (splitting each peak of the quartet into a doublet). The four closely spaced quartets at high field arise from the proton (β), the fine structure consists of a doublet splitting by (α), each component is then further split into a doublet by coupling with the CHO group proton and, finally, each component is split into a quartet by coupling to the methyl group.

(b) Shows the simplification in the spectrum that is found when the sample is irradiated at the frequency of the methyl group while examining the spectrum of the olefinic protons. The fine structure now shows no signs of the quartet splittings observed in (a)

is visible then a decoupling experiment may be used to give its chemical shift. There are two distinct ways of performing this kind of experiment, and each of these may be done in either the field- or the frequency-sweep mode.

In one method the unobscured resonance is observed and the decoupling frequency which gives optimal removal of fine structure is sought. An experiment of this type done in the field-sweep mode is illustrated[4] in Figure 3.8. The compound is the steroid 3,3-dimethoxy-2β,19-epoxy-5α-androstan-17β-ol

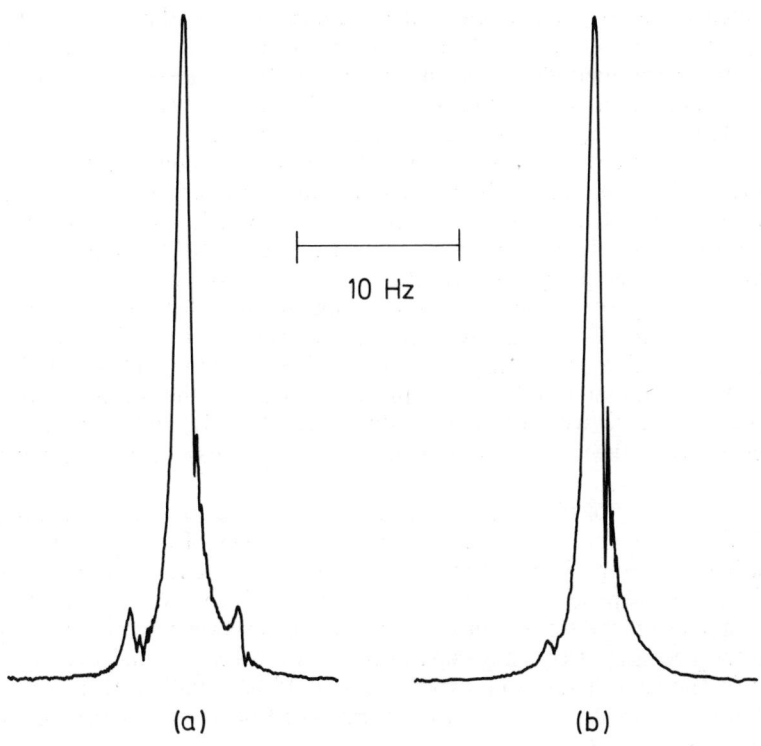

10 Hz

(a)   (b)

Figure 3.7. Identification of satellite lines by heteronuclear double resonance. (a) Proton spectrum of hexamethyl disiloxane, $(Me_3Si)_2O$, showing weak ^{29}Si (abundance 4.7%) satellites flanking the main peak. (b) Removal of the satellites by simultaneous irradiation at the ^{29}Si resonance frequency. Note that a very weak impurity peak becomes visible in the decoupled spectrum

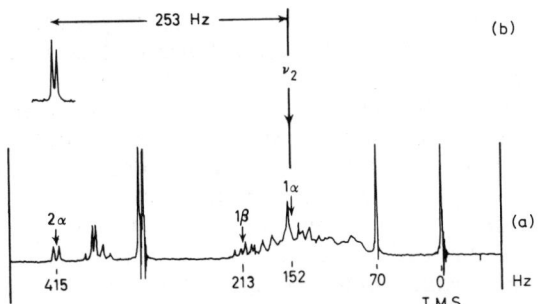

Figure 3.8. 100 MHz proton spectra of the steroid. (a) Normal spectrum in which the resonance of H(1α) cannot be discerned beneath the general methylene envelope. (b) Removal of the narrower splitting of the H(2α) resonance (arising from coupling to H(1α)) by irradiating at a point 253 Hz to high field of the H(2α) resonance. This gives the position of the H(1α) resonance. (Courtesy of American Chemical Society)

in which it was desired to know the chemical shift of the proton 1α for certain comparative studies. The resonance of this proton is hidden in the envelope of the methylene resonances in the 100 MHz spectrum, but the resonances of 2α and 1β are clearly visible 415 and 213 Hz respectively to low field of the reference tetramethylsilane (T.M.S.), see p. 6. The resonance of 2α is actually a poorly resolved double doublet, and consideration of the structure of the molecule shows that this fine structure must arise from coupling to the two protons attached to C_1. A field-sweep decoupling experiment in which the decoupling frequency was set 202 ($= 415 - 213$) Hz to high field of the point of observation showed that the wider splitting was due to coupling to 1β, and a similar experiment with a frequency difference of 253 Hz removed the narrower splitting as shown in (b). The chemical shift of the 1α proton is thus $415 - 253 = 162$ Hz to low field of T.M.S. at 100 MHz. A similar result was obtained by observing the resonance of 1β and decoupling 1α with a frequency difference of 51 Hz, and both of these experiments could have been performed with equal ease in the frequency-sweep mode.

The other way of obtaining these results would be to irradiate continuously the resonance of 1β or 2α while the remainder of the spectrum was recorded. A change in the appearance of the methylene envelope would then indicate the position of the 1α resonance. Clearly this type of experiment is most conveniently performed in the frequency-sweep mode, although once the necessary frequency separation of the observing and decoupling fields is known, a field-sweep experiment would be equally satisfactory. In all doubtful cases it is desirable to confirm assignments by performing both types of experiment.

Decoupling is just one of a wide range of double resonance experiments that may be performed, and details of applications of others will be given in Chapter 6.

Paramagnetic compounds

The presence of a paramagnetic compound normally causes broadening of n.m.r. spectral lines. In favourable cases, however, this broadening is not serious and useful results can be obtained. There have been several review articles[5] dealing with the n.m.r. spectra of paramagnetic species and the information about the paramagnetic species that can be obtained from such spectra. In the present context, we are more concerned with the use of paramagnetic materials as aids to interpreting n.m.r. spectra of diamagnetic compounds. Addition of paramagnetic material to a diamagnetic sample may result in (a) the loss of multiplicity due to spin-spin interaction, or (b) changes in the chemical shifts; the latter of these two effects is at present the more important.

Small amounts of nickel and cobalt compounds have been added to samples to remove the effects of spin-spin coupling from the spectrum. This has been successful for certain organo-phosphorus compounds, where it has been possible to remove the effects of ^{31}P splitting in the proton spectrum[6].

A more important development in recent years[7] has been the use of paramagnetic lanthanide complexes as 'shift reagents'. An early example of such a reagent is tris(dipivalomethanato)europium (usually abbreviated to $Eu(DPM)_3$).

The lanthanide in such a complex can increase its coordination number by interaction with lone pair electrons from other species. When the lanthanide complex is added to a suitable compound, association can occur and consequently the n.m.r. chemical shifts of the compound are altered. The resulting change in shift differs from site to site in the molecule, so peaks that are close together in the spectrum obtained from the compound alone may be separate in the spectrum obtained after addition of the shift reagent. Large shift differences may be produced so the spectrum of the sample may also become more amenable to first-order analysis. Although most work involving shift reagents has been concerned with proton spectra, there have been a few reports[8, 9] of the effects of shift reagents on the spectra of other nuclei such as ^{13}C and ^{14}N.

A number of different lanthanide complexes have been investigated. Usually $Eu(DPM)_3$ produces shifts to low field while $Pr(DPM)_3$ generally gives shifts to high field. The Pr complex gives the greater shifts and also produces rather greater broadening. The tris (1,1,1,2,2,3,3-heptafluoro-7,7-dimethyl-4,6-octanedione) complexes $Eu(FOD)_3$ and $Pr(FOD)_3$ have the advantage of greater solubility in typical solvents. It has been reported[9] that $Dy(DPM)_3$ is the best high-field, and $Yb(DPM)_3$ is the best low-field reagent for ^{14}N.

As the induced shift changes are dependent on the amount of shift reagent added, it is customary to report values of the induced shift obtained by linear extrapolation to a molar ratio of 1 : 1. In addition the shifts are temperature dependent so it may be possible to increase the shifts by lowering the sample temperature.

Satellite peaks

As the common isotope of carbon, ^{12}C, has $I = 0$, the n.m.r. spectra of hydrogen nuclei in organic compounds show no fine structure due to spin-spin interaction with ^{12}C. There is, however, a 1% natural abundance of the isotope ^{13}C which has $I = \frac{1}{2}$, and this isotope can spin couple to give a doublet splitting in the n.m.r. spectra of other magnetic nuclei. If all the details of the hydrogen n.m.r. spectrum of a simple compound, such as chloroform are obtained, they will be found to consist of a single intense peak due to the species $^{12}CHCl_3$ flanked symmetrically by a weak doublet arising from the species $^{13}CHCl_3$. The separation between the components of the doublet is

$J_{13_{C-H}}$ = 209 Hz, and the intensity of each component of the doublet is about $\frac{1}{200}$ the intensity of the peak due to the more abundant species $^{12}CHCl_3$. The peaks of the low intensity doublet are referred to as ^{13}C satellites of the main peak. Satellite peaks can arise when an element exists in more than one isotopic form, one isotope having $I = 0$, the others having $I > 0$. Examples are:

(a) the hydrogen n.m.r. spectra of organotin compounds show two sets of satellites, one due to ^{117}Sn (abundance 7.67%) and the other due to ^{119}Sn (abundance 8.68%), the intensity of each satellite peak being about $\frac{1}{20}$ the intensity of the peak arising from $I = 0$ species

(b) the species ^{195}Pt ($I = \frac{1}{2}$, abundance 33.7%) can give rise to a doublet splitting in the hydrogen spectrum of organo platinum compounds. In this case each satellite peak has $\frac{1}{4}$ the intensity of the peak from species containing non-magnetic platinum isotopes. In the hydrogen n.m.r. spectrum each absorption appears as a $1:4:1$ triplet.

If X is the element existing in several isotopic forms and the n.m.r. spectrum from element A is observed, then the X satellites in the spectrum of A can assist in the solution of problems in the following ways:

(a) The coupling constant J_{A-X} can be measured from the separation of the satellite peaks. The magnitude of the coupling may give information about such things as hybridisation of X; the valency of X; the number of bonds between X and the hydrogen nucleus showing spin-spin interaction. Here care must be taken in case factors such as electronegativity of substituents are causing the coupling constant to take up a misleading value.

(b) Double resonance experiments (in which the spectrum of A is observed while the sample is irradiated at the frequency of X transitions) enable the spectrum of X to be obtained from the A spectrum and are described in Chapter 6.

(c) The presence of different isotopic species of X in a molecule may remove the equivalence of certain A nuclei. The proton magnetic resonance spectrum of each of the species (a) to (c) is a single peak;

(a)   (b)   (c)

this is insufficient to distinguish isomers. Satellite peaks arise from the species (d) to (f). For (d) the satellites[10] are a doublet with separation

(d)   (e)   (f)

$J_{13_{C-H}}$ = 166 Hz. For (e) and (f) further fine structure appears (Figure 3.9); this arises because of the splitting produced by the ^{13}C nuclei. The proton bonded to ^{13}C is no longer equivalent to the proton

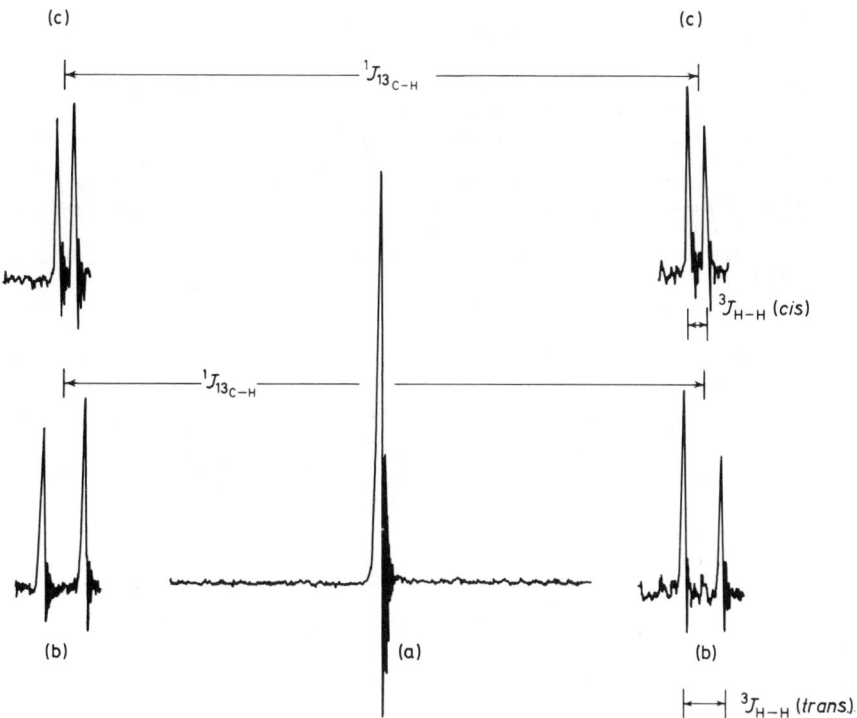

Figure 3.9. ^1H n.m.r. spectra of 1,2-dichloroethylenes. The peak (a) arises from ^{12}C—H in the *trans*-isomer; the gain of the instrument was reduced while this peak was scanned. The peaks (b) arise from ^{13}C—H species in the *trans*-isomer. The upper peaks (c) arise from ^{13}C—H in the *cis*-isomer. There is a chemical shift difference of about 0.04 p.p.m. between the two isomers

bonded to the ^{12}C atom so a splitting arises owing to coupling between the two hydrogen nuclei. The ^{13}C satellites in (e) and (f) each consist of a doublet of doublets in which the larger splitting is $^1J_{13_{C-H}}$ and the smaller is $^3J_{H-H}$ In ethylenic systems the coupling $^3J_{H-H}$ is greater for the *trans* arrangement of hydrogen nuclei than it is for the *cis* isomer, so the magnitude of the smaller coupling in the ^{13}C satellites of the ethylene dichlorides can be used to distinguish between (e) and (f).

References

1. VARIAN ASSOCIATES, *High Resolution N.m.r. Spectra*, **1** (1962), **2** (1963)
2. LASZLO, P., *Prog. in N.m.r. Spectro.*, **3**, 231 (1967)
   RONAYNE, J., and WILLIAMS, D. H., *Annual Rev. N.m.r. Spectro.*, **2**, 83 (1969)
3. HATTON, J. V., and RICHARDS, R. E., *Mol. Phys.*, **3**, 253 (1960)
4. BHACCA, N. S., and WILLIAMS, D. H., *Applications of N.m.r. Spectroscopy in Organic Chemistry*, Holden-Day (1966)
5. WEBB, G. A., *Annual Reports of N.m.r. Spectro.*, **3**, 211 (1970) and references therein
6. KAINOSHO, M., *J. Phys. Chem.*, **73**, 3516 (1969)
   ENGEL, R., *Chem. Commun.*, 133 (1970)
7. HINCKLEY, C. C., *J. Amer. Chem. Soc.*, **91**, 5160 (1969)
   SANDERS, J. K. M., and WILLIAMS, D. H., *Chem. Commun.*, 422 (1970)
   BRIGGS, J., FROST, G. H., HART, F. A., MOSS, G. P., and STANIFORTH, M. L., *Chem. Commun.*, 749 (1970)
8. BRIGGS, J., HART, F. A., MOSS, G. P., and RANDALL, E. W., *Chem. Commun.*, 364 (1971)
9. WITANOWSKI, M., STEFANIAK, L., JANUSZEWSKI, H., and WOLKOWSKI, Z. W., *Chem. Commun.*, 1573 (1971)
10. WHIPPLE, E. B., STEWART, W. E., REDDY, G. S., and GOLDSTEIN, J. H., *J. Chem. Phys.*, **34**, 2136 (1961)

4 The n.m.r. spectrometer

The essentials of a high resolution nuclear magnetic resonance spectrometer are:

(a) a very stable homogeneous magnetic field together with a means of varying this field over a small range in a controlled manner
(b) a stable source of radio-frequency power
(c) a detection and display system.

A block diagram showing the main features of an n.m.r. spectrometer is given in Figure 4.1.

4.1 The magnetic field

A suitable magnetic field of the required stability and homogeneity may be provided by either a permanent or an electromagnet. The advantages of the former are:

low power consumption, the only power required for the magnet is that needed to operate the thermostat system maintaining the magnet at a constant temperature;
high inherent stability, once installed the magnet should always be ready for use, thus eliminating any warm-up period immediately after switching on.

Disadvantages are:

great sensitivity of field strength to changes in temperature, so that careful thermostatting (to better than 0.1 deg) of the entire magnet is essential;
the need to pre-heat samples to the magnet temperature;
sensitivity to movement of ferromagnetic bodies (such as gas cylinders, bunches of keys, elevators) in the vicinity of the magnet;
the need for a very narrow gap between the pole faces if high field strengths are to be obtained;
the inability to vary the field strength sufficiently to accommodate the spectra of different elements at a particular radio frequency.

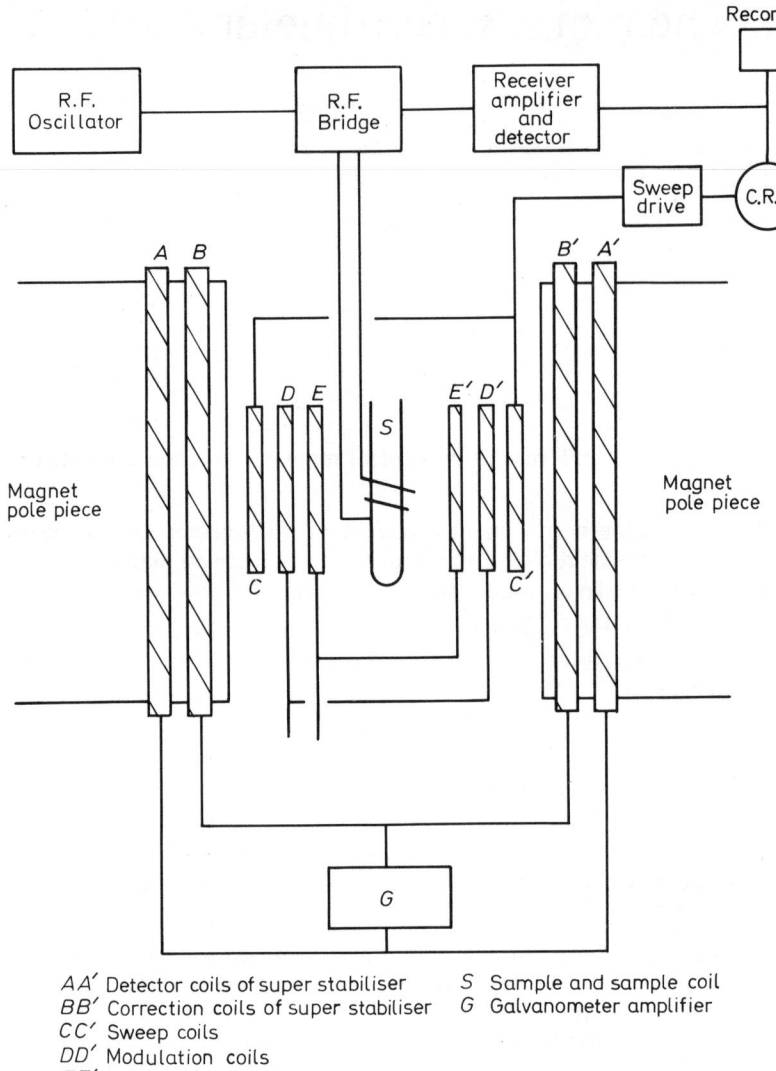

AA′ Detector coils of super stabiliser   S   Sample and sample coil
BB′ Correction coils of super stabiliser   G   Galvanometer amplifier
CC′ Sweep coils
DD′ Modulation coils
EE′ Shim coils

Figure 4.1. Block diagram of n.m.r. spectrometer

The last of these is not so serious a difficulty as was once supposed. The advantages of electromagnets are:

small effect of temperature upon field strength;
no need to pre-heat samples;
lower sensitivity to the effects of external ferromagnetic materials;
large gaps available at high field strengths;
ability to operate over a wide range of different field strengths.

Disadvantages are:

high consumption of electric power, leading to high operating costs and the generation of considerable heat; typically 5 kW of power will be needed, together with about 1 litre/minute of cooling water;

after switching on there is a significant warm-up time before best stability of the magnetic field is attained;

lower field stability due to sensitivity to fluctuations in the power supply.

In practice, electromagnets usually incorporate safety devices to guard against overheating or cooling water failure, but transient fluctuations in the power supply can make these operate unnecessarily with subsequent delay owing to the warm-up time.

Originally electromagnets were designed to operate at high voltage (2000–4000 volts) and low currents (1–2 amperes), but recently low impedance systems have been developed and these use solid state power supplies working at low voltage. Another development has been the introduction of super-conducting solenoids that can give fields of 5 tesla or more, with adequate homogeneity and stability for high resolution work[1]. As these solenoids operate at liquid helium temperature much auxiliary equipment is needed; at present such spectrometers are not common.

In practice it is found that none of the above systems gives a sufficiently stable magnetic field for high resolution n.m.r. and a device known as a field corrector, or flux stabiliser, or super stabiliser is incorporated in commercial instruments. This takes the form of a pair of coils (*AA'*, Figure 4.1) placed so that changes in the main magnetic field strength induce an e.m.f. in the coils. This e.m.f. is amplified (Galvanometer amplifier *G*, Figure 4.1) and used to control a current passing through a second set of coils (*BB'*, Figure 4.1) in such a way as to compensate exactly for the original change in field strength. The response of this type of system is quite rapid and the residual magnetic field fluctuation can be reduced to less than 1 in 10^8 by these means; however, a very slow overall drift of the main field cannot be corrected in this way. More sophisticated devices which can maintain the magnetic field at a predetermined value with high precision for an indefinite period by using the n.m.r. phenomenon itself will be described later.

In addition to having high stability, the magnetic field needs to be uniform over the volume of the sample. If this is not so the lines of the recorded spectrum will be excessively broad as shown in Figure 4.2, and narrow splittings will not be resolved.

By careful design both permanent and electromagnets can achieve residual inhomogeneities of as little as 1 in 10^7 throughout a 0.5-ml sample. This can be improved (Figure 4.2) by spinning the sample about an axis (usually the vertical axis, perpendicular to the field direction) at about 2000 rev./min. This helps to average out field gradients along the two other axes. Further improvement can be achieved by the use of shim (or Golay) coils[2] (*EE'*, Figure 4.1). These are coils mounted on either the probe or the pole faces of the magnet, designed to produce a weak magnetic field having a gradient that can be varied by altering the current passing through the coil. The shim coil currents are adjusted to produce a field gradient at the sample which cancels any gradient in the field of the laboratory magnet. The use of a standard sample to set up shim coils is discussed in Section 5.2. The number of sets of shim coils used in different instruments varies from four to ten; by use of them the residual inhomogeneity can be reduced to a few parts in 10^9. In practice it is found that the only shim coil to need frequent adjustment is the one which controls field gradients along the axis of spinning (the y axis), and devices are now available to perform this operation automatically.

The volume over which satisfactory homogeneity can be obtained limits the size of sample that may be used. For high resolution work with ^{1}H, ^{19}F, and ^{31}P, cylindrical samples 5 mm in diameter are suitable, but for other isotopes where low inherent sensitivity causes problems it is common to use

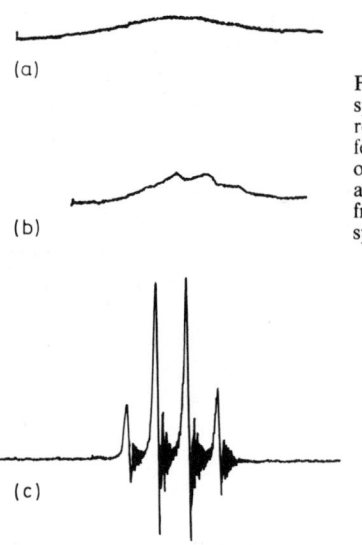

(a)

(b)

(c)

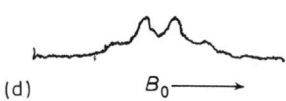

(d) $B_0 \longrightarrow$

Figure 4.2. Appearance of the quartet in the ^{1}H n.m.r. spectrum of CH_3CHCl_2 under different conditions of resolution. The spectra were run within a period of a few minutes; the r.f. power-level and the gain settings of amplifiers were the same throughout. Differences of appearance and peak height among the spectra arise from differences in resolution. The figures show the spectra obtained when the sample was:

(a) in an inhomogeneous field, not spinning; no fine structure is resolved and the absorption is so broad that it is almost undetected;

(b) in the same field as (a), spinning; although the lines are still broad they are slightly sharper (and of greater height) than in (a), traces of fine structure are discernible;

(c) spinning in a homogeneous field obtained by using shim coils adjusted to compensate for field gradients at the sample; the fine structure is well resolved, peak heights have increased as the line-widths have narrowed;

(d) in the same field as (c), not spinning; although fine structure is detectable it is not as well resolved as in (c), the line-width is greater and the peak height has diminished

larger samples, of up to 15 mm diameter. Samples larger than about 9 mm diameter are not usually suitable for spinning. The loss in resolution resulting from the use of large samples is not as serious as might be supposed. Most of these other nuclei have resonance frequencies lower than that of the proton, and a given loss of *absolute* homogeneity will lead to a correspondingly smaller increase in line-width. Furthermore, these other nuclei are often associated with large chemical shift ranges and coupling constants, so the requirements with regard to line-widths are not so stringent.

Overall variation of the resultant magnetic field strength at the sample can be accomplished by passing a suitable current through a pair of Helmholtz coils (the sweep coils, *CC′*, Figure 4.1) mounted in the gap of the magnet. Alternatively, an artificial error signal can be fed into the super-stabiliser control system, this will then produce a correcting field change and so generate the required sweep. This generally offers a more precise means of achieving controlled variation of the magnetic field strength.

4.2 Radio-frequency circuits[3]

Radio-frequency power supply

A low power ($\sim$ 1 watt) source of radio-frequency energy, with a frequency stability of 1 in 10^9 is required. This is conveniently derived from a quartz crystal controlled oscillator. The crystal itself should be protected from rapid changes of temperature. It turns out that slow, long-term, drifts in

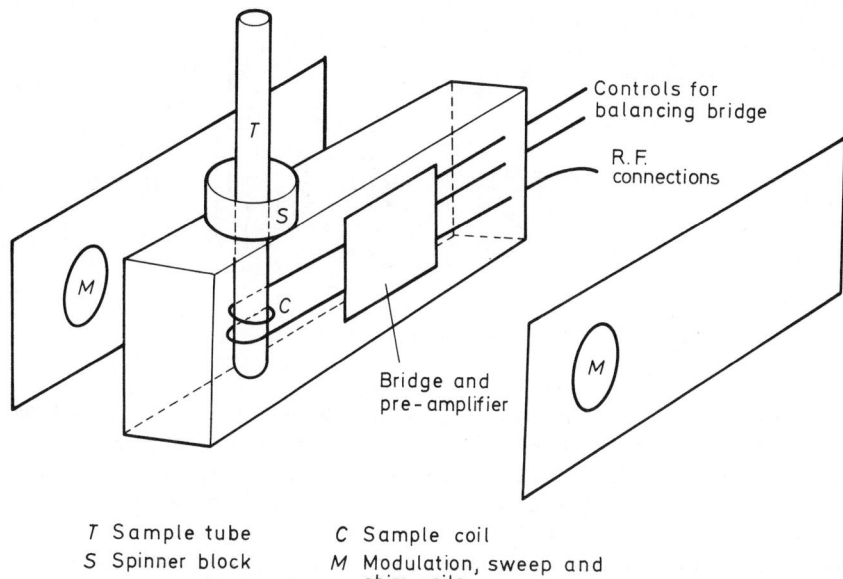

T Sample tube   C Sample coil
S Spinner block   M Modulation, sweep and shim coils

Figure 4.3 (a). Exploded view of probe unit for a single coil instrument

frequency are not important, and many manufacturers have taken advantage of this to simplify the thermostatting of the quartz crystals in their spectrometers. Means of controlling the power output of the oscillator (or transmitter) are provided, while facilities for modulating it with an audio frequency are also often included.

The detection system

This is one of the most important parts of an n.m.r. spectrometer as the ultimate signal-to-noise ratio that can be attained depends on the detection system. The two types of detection in common use are:

Single coil arrangement. Energy from the radio-frequency power supply is fed to a coil (the sample coil, Figure 4.3 (a)) wound about the sample. The coil forms part of a radio-frequency bridge circuit; energy absorption by the sample produces changes in the balance of the bridge which are detected by the receiver.

Crossed coil arrangement. In this method use is made of two coils, these are arranged with their axes perpendicular to one another and also to the magnetic field. Energy from the r.f. oscillator is fed to the sample via the transmitter coil. When the sample absorbs energy an e.m.f. is induced in the second (sample or receiver) coil and this can be detected by the receiver.

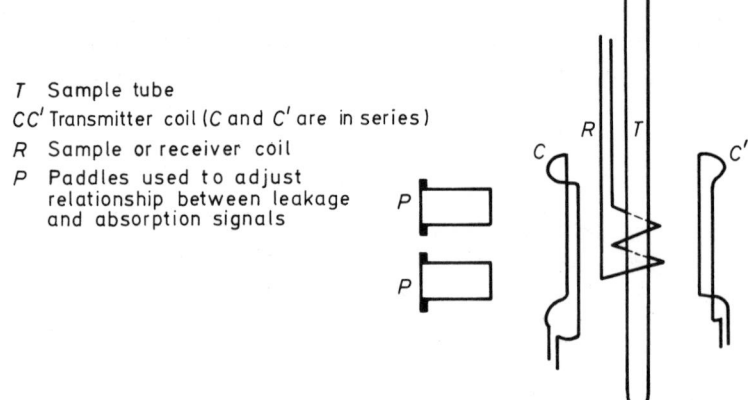

T   Sample tube
CC′ Transmitter coil (*C* and *C′* are in series)
R   Sample or receiver coil
P   Paddles used to adjust relationship between leakage and absorption signals

Figure 4.3 (b). Coil arrangement for nuclear induction instrument

This is often known as the nuclear induction method. The coils are mounted in the *probe unit*; Figure 4.3 shows the essential features of probes. In practice, the receiver will detect a direct or leakage signal from the transmitter independently of any absorption by the sample, use is made of the leakage signal in the detection process. The relationship between this signal and that due to n.m.r. absorption affects the shape of the recorded resonance. It may be altered by devices incorporated either in the detector circuitry or, for a crossed coil instrument, in the probe, and this adjustment will be dealt with later in Section 5.2 under the heading Phase Adjustment.

The signals to be detected are very weak, so the receiver must have a high sensitivity and care is needed to reduce spurious signals (noise) to a minimum. Noise may be the normal electrical noise that occurs in all circuits or it may be generated mechanically by, for example, the spinning of the sample. To obtain the best signal-to-noise ratio the first stage of amplification of the signal takes place in a pre-amplifier unit, which is normally incorporated in the probe itself, as close as possible to the receiver coil. The transmitter and receiver proper are usually built into a single unit and in addition to the control for the power output of the transmitter a means of altering the gain of the receiver is also provided. The time constant of the output circuit of the receiver is usually variable so that the operator can vary time constant and sweep rate to obtain the most favourable signal-to-noise ratio for each sample. The output from the receiver is a voltage, which is fed to a chart recorder or to the Y plates of a cathode ray oscilloscope. In the latter case the spectrum is displayed by applying a suitable sawtooth voltage to the X plates of the oscilloscope and to the sweep coils of the spectrometer. The magnetic field is thus swept repetitively over a small (adjustable) range in step with the horizontal movement of the spot on the oscilloscope screen. Not all spectrometers have this facility for oscilloscope display of the signal.

4.3 The uses of modulation

Originally modulation was used in n.m.r. spectroscopy as a means of calibrating spectra; subsequently a number of improvements to spectrometers have been achieved by application of effects dependent on modulation. For present purposes it will be sufficient to consider modulation simply as a process which mixes signals of different frequencies. The convention will be adopted that the symbol v will be given to frequencies in the megahertz (or radio-frequency) region while audio frequencies will be given the symbol f. When a radio frequency v is modulated by an audio frequency f the resulting signal can be considered as being made up from the frequencies $v-f$; v; and $v+f$. If the intensity of the audio-frequency field becomes sufficiently great it may be necessary to consider the resultant signal as being made up from the frequencies $v-2f$; $v-f$; v; $v+f$; and $v+2f$.

Calibration

A disadvantage of the simple spectrometer system outlined above is that no constant relationship will exist between the change in the magnetic field during a spectral sweep, and the distance traversed in the x direction of the chart paper. Each spectrum must be individually calibrated, and this can be achieved by modulating the output of the radio-frequency transmitter with a suitable audio-frequency signal. Rather than modulating the radio-frequency signal, it is often more convenient to modulate the magnetic field by applying the modulation frequency to extra coils (DD', Figure 4.1) in the magnet gap. This produces a similar effect to modulation of the radio-frequency signal.

Figure 4.4 shows that modulation results in generation of sidebands in the spectrum; these are spaced at a distance proportional to the modulation frequency, so that interpolation between peaks on the recorder chart will yield accurate line separations.

Frequency-sweep spectra

The most convenient way of scanning frequency to produce frequency-sweep spectra is achieved by use of modulation effects. This arises because of the practical difficulties associated with producing a variable radio-frequency oscillator having adequate frequency stability, together with the means of varying this frequency over a small range in a controlled manner. If, for example, we accept 0.6 Hz as a criterion of acceptability, then an oscillator operating at 60 MHz would have to be stable to 1 part in 10^8 and variable with the same precision. The requirements for an audio-frequency oscillator

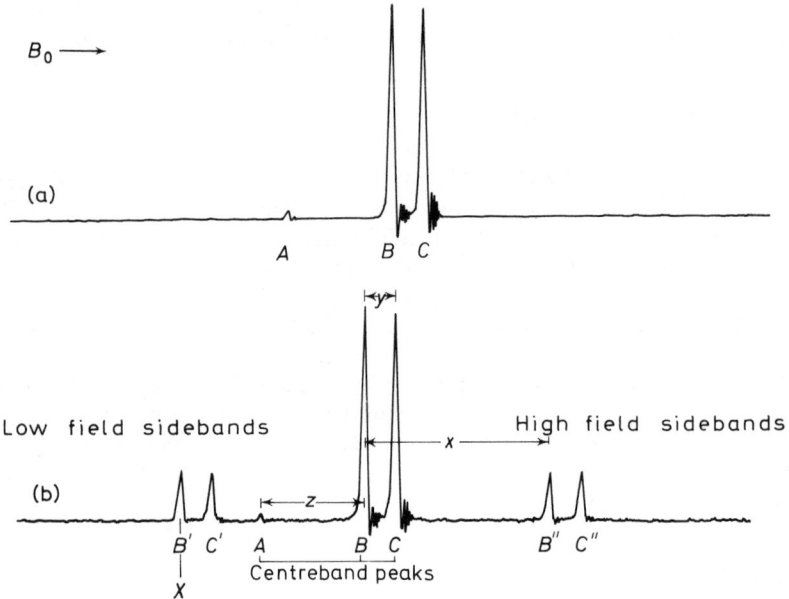

Figure. 4.4. Modulation effects in n.m.r. spectra.

(a) The spectrum of a sample in which no modulation has been applied.

(b) The spectrum of the same sample, the magnetic field here being modulated at an audio frequency $f_m = 60$ Hz. B' and C' are the lower (or low-field) while B'' and C'' are the upper (or high-field) sidebands of the original peaks B and C (the centreband peaks). The strength of the modulation signal is weak enough for the sideband peaks to be of lower intensity than the centreband peaks. The peak A is so weak that under these conditions its sidebands cannot be detected.

  In spectrum (b) the distance x cm (which should be the same for the separations B to B', B to B'', C to C', and C to C'') is equivalent to 60 Hz. Then y cm, the separation between peaks B and C is equivalent to

$$\frac{y \cdot 60}{x} \text{ Hz}$$

which in this case works out to be 10 Hz. Similarly the frequency separation between peaks A and B is

$$\frac{z \cdot 60}{x} = 33 \text{ Hz}$$

used to modulate a radio-frequency signal are less rigorous; a stability of 0.6 Hz corresponds to 1 part in 10^2 at 60 Hz, or 1 part in 10^4 at 6 kHz. By combining a variable audio-frequency with a fixed radio-frequency oscillator (a crystal-controlled fixed radio-frequency oscillator of required stability being readily attainable) the sideband frequencies $v-f$ and $v+f$ have the required stability plus variability making them suitable for generating frequency-sweep spectra. Figure 4.4 (b) shows the field-sweep spectrum obtained using a modulation frequency of 60 Hz. If the magnetic field were set to X, corresponding to the position of B' the low-field 60 Hz sideband, the spectrometer would indicate an absorption of energy by the sample. When the modulation frequency f is reduced to, say, 50 Hz there will be no absorption of energy by the sample. If the modulation frequency is now varied continuously, no absorption of energy will be detected until $f = 60$ Hz, at which stage the B'' sideband peak will be displayed. If f is varied further C'' (the sideband of peak C) will be detected when $f = 60 + 10 = 70$ Hz (the

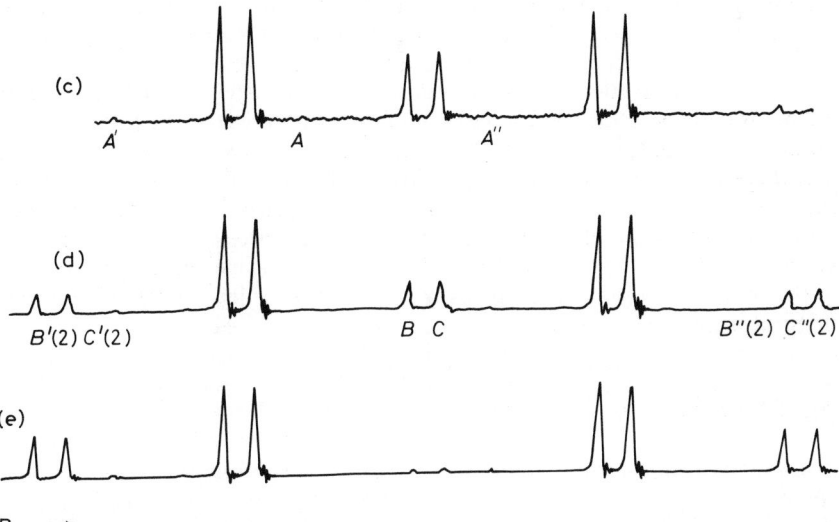

$B_0 \longrightarrow$

Figure 4.4. *continued*
  (c) This spectrum shows the effect of increasing the strength of the modulation signal. The intensity of the sidebands has increased (A' and A'' the sidebands of peak A are just detectable) while the intensity of the centreband peaks has decreased. In the centreband spectrum peak A is scarcely detectable.
  (d) Further increase in the strength of the modulation field leads to the appearance of second sidebands $B'(2)$, $B''(2)$, $C'(2)$, and $C''(2)$ with a further diminution in the intensity of the centreband peaks. The separation between a centreband peak and its second sideband (e.g. between B and $B'(2)$) is equal to twice the modulation frequency f_m.
  (e) The intensity of the modulation field has here been increased to such an extent that the centreband has almost disappeared.

separation between B and C being 10 Hz). In this illustration the modulation frequencies used have been chosen for convenience in producing reasonable diagrams. As proton chemical shifts at 60 MHz extend over a range of about 800 Hz, a modulation frequency of a few kHz would be used in practice to avoid difficulties arising from overlap of centre and sideband peaks. The present illustration has described the use of the lower sideband; it is equally possible for the upper sideband to be used for this purpose and different manufacturers may make use of different sidebands. For the interpretation of some double resonance experiments (see Chapter 6) it may be necessary to know whether the upper or lower sideband is being used.

Baseline stabilisation

A further disadvantage of the simple spectrometer system is that small extraneous changes of conditions (such as temperature) in the probe will alter the strength of the signal detected by the receiver, and consequently the baseline will tend to fluctuate or drift. This can be overcome by using a device known as a phase-sensitive detector (p.s.d.[4]) which depends on modulation for its operation. The radio-frequency signal or the magnetic field is modulated with a fixed audio frequency f_B of a few kilohertz; this frequency f_B is also fed directly to the p.s.d. This then ensures that only signals at the frequency of one of the sidebands and with the correct phase relationship are detected by the receiver. Overall changes in r.f. power level will then have no effect and a stable baseline is achieved.

Baseline stabilisation by this method makes it extremely easy to record an integrated spectrum, that is a graph of the summation of the total spectral intensity. The output from the audio-frequency phase-sensitive detector is a voltage which can be transformed into a current. As a peak is traversed, the total electric charge carried by the current is proportional to the area of the peak, and this charge can be stored on the plates of a capacitor. The voltage produced across the terminals of the capacitor is also proportional to the peak area, and it can be transferred to the recorder to give the required integral trace.

Field-frequency locking systems

The inherent long-term drifts of field strength which remain particularly in spectrometers equipped with electromagnets can be eliminated by use of the n.m.r. phenomenon itself. A control sample of high proton content (usually water) is built into the probe as close as possible to the experimental sample. The control sample is provided with its own n.m.r. circuitry and gives rise to a resonance signal in the usual way. Any change in the intensity or phase of this signal (indicating a drift in the field strength) is used to actuate an electronic feed-back loop which restores the field strength to its original value. In this way the field strength can be held constant to 1 in 10^8 indefinitely. Strictly speaking it is not the strength of the field in absolute terms that must be held constant, but rather it is the ratio of field strength to frequency of irradiation signal that should not vary (see Eq. (2.4)). The circuit mentioned above does not distinguish between changes in field and changes in frequency so it will, in fact, correct for both drifts in field and drifts in frequency. This stabilisation arrangement is generally described as an external field-frequency lock, because the control sample is separate from the experimental sample. This type of locking system is particularly convenient for routine work because the field remains locked when the experimental sample is changed; it is rare for drifts to exceed 0.5 Hz. In addition, the water control sample can be used for locking the field when the resonances of nuclei other than the proton are observed.

External locking systems depend on the assumption that changes in the field strength at the experimental sample are paralleled by changes at the control sample. This will not be exactly true since the two samples are normally separated by a few centimetres and, for this reason, *internal* locking systems have been devised. In these, *two* separate audio frequencies (f_1 the locking frequency and f_2 the observing frequency) are used to modulate the radio frequency (or alternatively the magnetic field) and the r.f. energy spectrum is then as shown in Figure 4.5. The field strength is adjusted to X_1 corresponding to the f_1 sideband of a sharp line in the spectrum of the

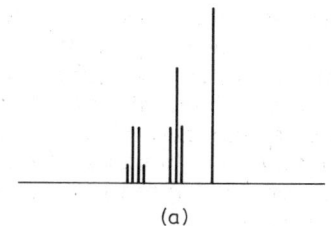

(a)

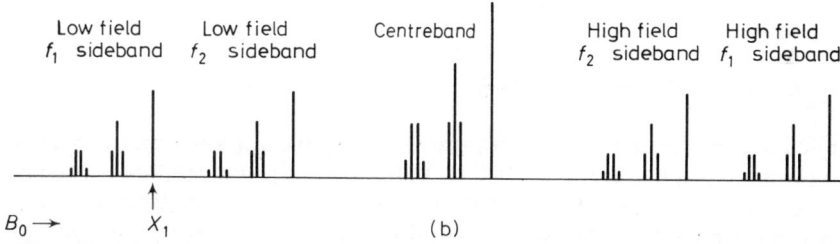

| Low field f_1 sideband | Low field f_2 sideband | Centreband | High field f_2 sideband | High field f_1 sideband |

$B_0 \longrightarrow$ X_1 (b)

Figure 4.5. Diagrammatic representation of field-sweep spectra obtained (a) without modulation; (b) with modulation at two frequencies (f_1 and f_2), f_1 being of higher frequency than f_2. The relative intensities of the sidebands could be varied independently by altering the strengths of the modulating signals. In this example the text describes the use of the low-field sideband for locking; it is, of course, equally possible to use the high-field sideband as a lock signal

sample being examined and the r.f. detector output is passed through a phase-sensitive detector operating at a frequency f_1. The output from this p.s.d. is then used to actuate a control loop to the flux stabiliser, that maintains the ratio of field strength to frequency constant at a value governed by the relation

$$\frac{v - f_1}{B_0} = \frac{\gamma(1 - \sigma)}{2\pi} \qquad (4.1)$$

In this relation v is the frequency of the r.f. oscillator, and the lower sideband f_1 is used for locking; σ is the shielding constant of the nucleus giving the locking signal. Other resonances from the sample can now be detected by varying the second modulation frequency f_2 through a suitable range, and taking the output of the r.f. detector to a second p.s.d. operated at the frequency f_2. The output of this p.s.d. is then fed to the recorder or to the oscilloscope. It will be noticed that in this mode of operation the main magnetic field remains constant throughout, so this is a true frequency-sweep experiment. The stability achieved depends upon the sharpness of the line chosen to provide the locking signal, and upon the frequency stability of the oscillator used to generate f_1. Typically, the drift over several hours will be less than 0.1 Hz.

If a field-sweep spectrum with internal lock is required then the observing frequency f_2 may be kept constant, while the locking frequency f_1 is varied. This changes the frequency of the locking signal, which gives rise to a compensating change in the field strength in accordance with Eq. (4.1). This mode of operation is slightly less satisfactory, in that attempts to perform very rapid sweeps may cause the locking to be lost. The requirements of frequency stability for the oscillators generating f_1 and f_2 are not specially high (say 2 in 10^5); if the frequency drive of the variable-frequency oscillator is coupled to the drive of an $X-Y$ recorder, precalibrated charts may be used. Alternatively, individual spectra may be calibrated by measuring f_1 or f_2 as appropriate. The addition of a third audio-frequency oscillator f_3 makes it possible to generate a sideband at v_3 which can be used for homonuclear decoupling experiments (see p. 71).

4.4   Other accessories

Variable temperature attachment

Facilities for varying the temperature of the sample are commonly provided on commercial n.m.r. spectrometers. Variation of the temperature is usually achieved by passing a stream of air or nitrogen at the required temperature past the sample tube. The stream of hot or cold gas is transferred in and out of the probe through a dewar system so that the magnet is protected from temperature changes. Temperatures above ambient may be attained by passing the air or nitrogen over an electrically heated nichrome spiral. A thermocouple placed in the gas stream close to the sample gives an indication of the actual temperature of the sample and can also be used to operate a proportional control system that regulates the current to the heater and/or the gas flow rate, so maintaining the sample temperature constant to about ± 1 deg. Temperatures below ambient are attained either by using a stream of cold nitrogen from a liquid nitrogen boiler (in which case the heater evaporating the nitrogen can be regulated by the proportional control system), or by passing dry gas through a spiral metal tube immersed in a dewar containing liquid air. In the latter case, the temperature of the sample could be adjusted by altering the gas flow rate. When a sample temperature just below ambient is required, it is frequently found that temperature control is unsatisfactory if a stream of cold gas is used. This can arise because the gas flow rate required to produce the desired temperature is too low to give satisfactory regulation. In these circumstances the performance may be improved by using a higher flow rate (which would normally lower the temperature of the sample) and passing the stream of cold gas over the heater before it reaches the sample.

A recent innovation by one manufacturer enables low temperatures to be achieved by means of a gas expansion system located inside the probe. This eliminates the need for the external heat exchanger and the supply of liquid nitrogen. This system can be used down to $-50\,^{\circ}\text{C}$ with nitrogen or to $-100\,^{\circ}\text{C}$ with argon.

Double resonance facilities

Increasingly, manufacturers are providing the apparatus for homonuclear double resonance (see Section 5.8) as a standard part of their n.m.r. spectrometers, though it should be remembered that for some instruments this facility is an optional extra. The main additional instrumentation necessary for this type of experiment is a stable variable-frequency oscillator. The apparatus for heteronuclear double resonance almost invariably has to be obtained as an additional accessory for the spectrometer.

Spectral accumulation

One technique for enhancing the sensitivity of an n.m.r. spectrometer makes use of a computer of average transients (c.a.t.). The spectrum is scanned many times and the output of the spectrometer is fed into the c.a.t. Successive synchronised scans of the spectrum lead to reinforcement of the required positive signal, while random noise (either positive or negative) tends to be averaged out; this leads to improvement of the signal-to-noise ratio by a factor of $\sqrt{n}$, where n is the number of scans of the spectrum.

Automatic shim control

The need for long-term stability of resolution when a c.a.t. is being used has led to the introduction of *autoshim* devices. The autoshim is essentially a servomechanism which monitors a control signal from the sample, and adjusts the current through the shim coil regulating the y axis field gradient, so as to maintain maximum height of the control signal.

4.5 Commercial n.m.r. spectrometers

Table 4.1 summarises the features of a number of commercial n.m.r. spectrometers, on the basis of information obtained from the information sheets issued by their respective manufacturers. Models are continually changing, and new models and accessories are still being introduced; this table is up-to-date at the time of publication. Prices have not been included as these are continuously changing and are, in any event open to negotiation between the parties concerned.

The sensitivity (signal-to-noise ratio) and resolution claimed by manufacturers is continually being improved. They vary from instrument to instrument even for a given model, usually being appreciably better than specification. For current instruments resolution is better than 0.35 Hz, and the sensitivity (using the signal-to-noise ratio for the largest peak of the quartet in a 1% solution of ethyl benzene) is better than 30 : 1 for 60 MHz, and better than 50 : 1 for 100 MHz spectrometers.

Table 4.1 FEATURES OF COMMERCIAL N.M.R. SPECTROMETERS

Bruker: HFX 90: Electromagnet, field variable over full range; ^{1}H at 30, 60 and 90 MHz; probes available for ^{19}F, ^{13}C, ^{31}P and most other nuclei; internal homo- and heteronuclear locking facilities; field- and frequency-sweep recording; homo- and heteronuclear decoupling facilities; sample sizes 5–15 mm diameter; variable temperature range -170 to $+300\,°C$; F.T. (Fourier Transform) and time-sharing accessories available.

JEOL: C–60–HL: Electromagnet, limited range of field variation; ^{1}H at 60 MHz, ^{19}F at 56.4 MHz using same probe; probes available for other nuclei; internal and external locking facilities; field- and frequency-sweep recording; homo- and heteronuclear decoupling facilities; sample size 5 mm diameter (larger available); variable temperature range -150 to $+200\,°C$; F.T. accessory available.
  MH–100: Electromagnet, fixed field; ^{1}H at 100 MHz; field-sweep (ext. lock) frequency-sweep (int. lock); homonuclear decoupling; sample size 5 mm diameter; F.T. accessory available.
  PS–100: Electromagnet, wide range of field variation; ^{1}H at 100 MHz, ^{19}F at 94.1 MHz using same probe; probes for most other nuclei available; other details as for C–60–HL.

Perkin Elmer: R 12 B: Permanent magnet; ^{1}H at 60 MHz, ^{19}F at 56.4 MHz; field-sweep and frequency-sweep (int. lock) recording; homonuclear decoupling; sample size 5 mm diameter; variable temperature -100 to $+150\,°C$.
  R 20 B: Permanent magnet; ^{1}H at 60 MHz; probes available for ^{19}F, ^{13}C, ^{11}B and ^{31}P; field-sweep and frequency-sweep (ext. lock) recording; homo- and heteronuclear decoupling facilities; sample sizes 5–8 mm diameter; variable temperature -120 to $+200\,°C$.
  R 24: Permanent magnet; ^{1}H at 60 MHz; field-sweep recording; sample size 4 mm diameter.

Varian: T 60: Permanent magnet; ^{1}H at 60 MHz; probes available for ^{19}F, ^{13}C, ^{31}P, etc.; field-sweep recording (other modes as extras); homo- and heteronuclear decoupling; sample size 5 mm; variable temperature -100 to $+100\,°C$.
  XL–100: Electromagnet, field variable over wide range; ^{1}H at 100 MHz; probes available for most other nuclei; internal locking (homo- and heteronuclear), external locking as extra; field- and frequency-sweep recording; homo- and heteronuclear decoupling facilities; sample size 5–12 mm diameter; variable temperature -150 to $+200\,°C$; F.T. accessory available.

References

1. BECCONSALL, J. K., and MCIVOR, M. C., *Chem. in Britain*, **5**, 147 (1969)
2. ANDERSON, W. J., *N.m.r. and E.p.r. Spectroscopy*, Staff of Varian Associates, Pergamon, 176 (1960)
3. RADIO SOCIETY OF GREAT BRITAIN, *The Radio Communication Handbook* (1968)
    DONALDSON, P. K., *Electronic Apparatus for Biological Research*, Butterworth (1958)
4. ANDREW, E. R., *Nuclear Magnetic Resonance*, Cambridge, **44** (1955)
    FAULKNER, E. A., and STANNETT, R. H. O., *Electronic Engineering*, **36**, 159 (1964)

5 Experimental techniques

Throughout this chapter we assume that the spectrometer is functioning properly, and that the manufacturer's instructions with regard to preliminary adjustments have been complied with. The field homogeneity of an electromagnet depends on its past history, so in the course of preparing the instrument for use it may be necessary to carry out the procedure of *cycling* the magnet. This usually involves taking the magnet current above the required working value for a few minutes, then lowering the current until the desired field is attained. Cycling is particularly important for a magnet that is used at a number of field strengths for the study of different nuclei.

5.1 Preparation of sample

By comparison with many other spectroscopic techniques the quality of n.m.r. spectra is very dependent upon both spectrometer performance and sample quality. Even the sample container is important, and for the best results this should be carefully chosen. The sample tube must be of the proper diameter (commonly 5 mm outer diameter) and made from precision bore thin-walled tubing to maximise the amount of sample that can be accommodated. The shape of the bottom of the tube is quite critical, and the most uniform spinning together with freedom from spinning sidebands in the recorded spectrum is achieved only when the tube has a polished hemispherical base. Spinning sidebands are due to modulation of the magnetic field at the spinning frequency, and if their presence is suspected a useful check is to alter the spinning rate when their positions with respect to other peaks in the spectrum should change. Figure 5.1 shows the spinning sidebands that may occur with a poor sample tube.

Normally the sample tube can be closed with a plastic cap, but if a permanent all-glass seal is wanted, care must be taken to ensure that the tube remains evenly balanced. Tube requirements are much less stringent when large stationary samples are used (e.g. for ^{31}P or ^{13}C measurements in 13-mm tubes) and satisfactory results can often be obtained with ordinary test-tubes.

The sample itself must be a fluid, and will normally be a liquid of low viscosity; it should be free from suspended solid particles which may affect the resolution. If very small solid particles are present the centrifugal action of spinning may cause these to adhere to the walls of the tube, and the resolution may improve after some minutes in the probe. Alternatively, these particles may be removed conveniently by centrifuging the n.m.r. tube containing the solution, since the presence of a small amount of compacted solid at the bottom is not usually deleterious.

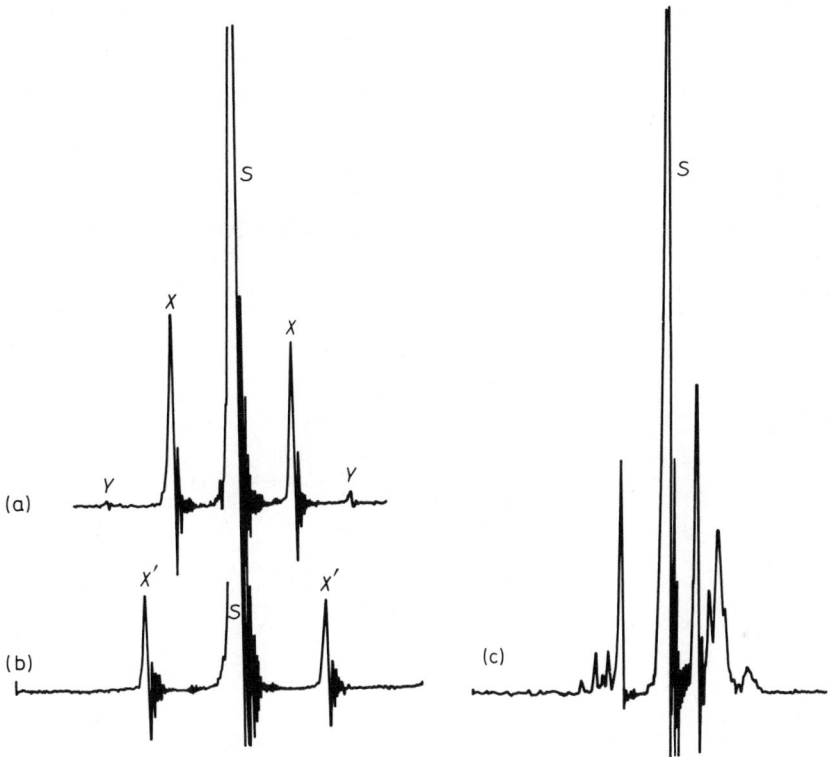

Figure 5.1. Spinning sidebands. In each example the spectrum should, ideally, be a single peak, S. Spectrum (a) has been superimposed on (b) (cutting off the top of peak S in (b)) so that the change in position of the sidebands can be seen more clearly. (a) Spinning sidebands XX are very pronounced, second sidebands, YY, are just detectable. (b) On using a faster spinning-rate than in (a) the sidebands X'X' move further away from the peak S, and also decrease in intensity. Second sidebands are not detectable. (c) In this example the tube was a poor fit in the spinner, in addition to the tube wobbling badly, the spinning-rate fluctuated

Even for samples which are themselves liquids it is desirable to use a solvent to minimise intermolecular interactions and radiation-damping effects. Ideally all data should be quoted for infinite dilution, this is not always convenient, so for many purposes a concentration of 5 mole% may be regarded as adequate dilution. For samples which are solids a solvent must be used, and its desirable properties include:

(a) Low viscosity to ensure complete averaging of the direct dipole-dipole interactions between the magnetic nuclei

(b) High solvent power
(c) Diamagnetism. Paramagnetic compounds usually broaden peaks in an n.m.r. spectrum
(d) Absence of any resonance signals which would obscure portions of interest in the spectrum of the sample. In practice, this means that a region of up to 80 Hz on each side of a strong solvent line may be excluded on account of spinning sidebands and ^{13}C satellites of the main peak. However, in spectrometers which have an internal field-frequency locking facility, it may be convenient to use the solvent resonance to provide a locking signal.
(e) Absence of any specific solvent-solute interactions, e.g. chemical reaction or the formation of hydrogen bonds.

Some commonly used solvents and their properties in these respects are given in Table 5.1.

The difficulties associated with interference from solvent peaks in proton spectra can be avoided by using deuteriated solvents. Many of these are now commercially available, and three popular, reasonably priced, ones are D_2O,

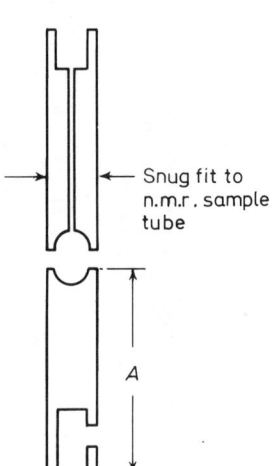

← Snug fit to
n.m.r. sample
tube

A

Figure 5.2. Microcell for use when only small amounts of sample are available. The two plugs are constructed from either nylon or P.T.F.E. *A*, the length of the lower component is such that the spherical cavity is level with the sample coil when the sample tube is supported in the spinner. The top component is drilled to facilitate filling. (Courtesy of The Chemical Society and Varian Associates)

$CDCl_3$, and $(CD_3)_2SO$. It should be remembered that deuteriated solvents contain a small percentage of the normal (protiated) species. Once a clear solution free from solid particles has been obtained the final precaution is to remove dissolved oxygen. This is necessary because molecular oxygen is paramagnetic and can lead to broad spectral lines. It is best done on a vacuum line using a repeated freeze–thaw technique; but not only is this often inconvenient it can also lead to tube breakages when certain solvents are used. Fortunately, it is normally adequate to pass a stream of nitrogen through the solution for a few minutes to flush out dissolved oxygen.

Some workers use a small ultrasonic cleaning bath for degassing samples. Some sort of treatment to remove dissolved oxygen is particularly important

Table 5.1 PROPERTIES OF SOME COMMONLY USED NON-DEUTERIATED N.M.R. SOLVENTS

Solvent	Liquid range (°C)	Volume diamagnetic susceptibility, $-K \times 10^6$ (20°C)	Performance with regard to			
			(a)	(b)	(d)	(e)
C_6H_{12}	+6 to 81	0.631	good	poor	singlet $\delta = 1.4$	good
CCl_4	−22 to 77	0.684	good	good	none	good
a$CHCl_3$	−63 to 61	0.740	good	very good	singlet $\delta = 7.27$	fair[b]
aC_6H_6	+6 to 78	0.626	good	good	singlet $\delta = 7.37$	poor
a$(CH_3)_2CO$	−95 to 56	0.460	good	good	singlet $\delta = 2.17$	poor
a$(CH_3)_2SO$	+18 to 189		bad	excellent	singlet $\delta = 2.62$	bad[c]
$(CH_3)_2NCHO$	−61 to 153		bad	excellent	lines at $\delta = 2.88$ $\delta = 2.97$ $\delta = 8.02$	bad
aCH_3CN	−45 to 80	0.534	good	good	singlet $\delta = 2.00$	fair
aCF_3COOH	−15 to 72		good	good	singlet variable usually to low field	bad
aH_2O	0 to 100	0.721	fair	good	singlet variable often about $\delta = 5$	bad
aC_5H_5N	−42 to 115	0.612	good	good	multiplets $\delta = 7.0$ $\delta = 7.6$ $\delta = 8.6$	poor
$C_4H_8O_2$ (1,4)	12 to 101	0.606 (at 32°C)	good	good	singlet $\delta = 3.0$	fair

(a) viscosity (b) solvent power (d) signals in ¹H n.m.r. spectrum (e) absence of solvent–solute interactions

a Obtainable as deuteriated compound from commercial sources
b Often contains a small amount of ethanol which may be removed by treatment with molecular sieves. Chloroform so treated should be stored in the dark to avoid the formation of phosgene
c Often contains water which will give an additional peak. The solvent may be dried over calcium hydride followed by distillation under reduced pressure

when aromatic compounds are present in the sample or the solvent; however, for much routine work it is often omitted.

It is important to ensure that the depth of the solution in the n.m.r. tube is adequate, for if the meniscus is in the region of the receiver coil the resolution will be impaired. Usually 3 cm is a sufficient depth, and special cells which incorporate movable Teflon or nylon plugs are available for small sample volumes. Figure 5.2 shows the Varian microcell which can be used with as little as 3×10^{-8} m^3 of liquid. Other designs are now available commercially.

5.2 Preparation of spectrometer

We deal here only with certain frequent adjustments that are likely to be common to all instruments.

Adjustment of resolution

This is usually done using a standard sample supplied with the spectrometer, although final trimming should be done on the actual sample to be examined. The number and arrangement of the shim coils vary considerably from one instrument to another, and the manufacturer's handbook should be consulted for details of the course of the adjustments. It is generally found

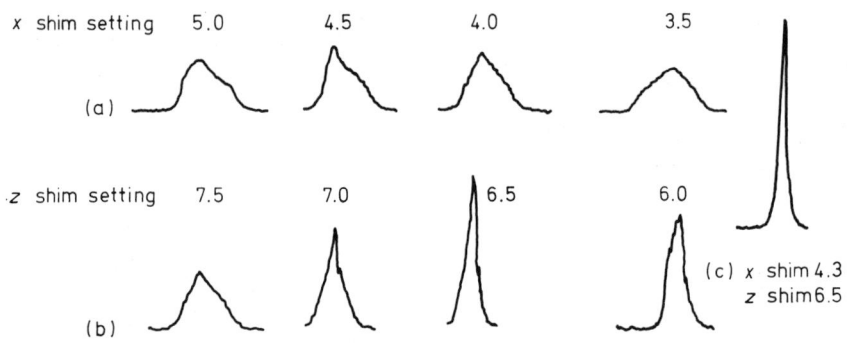

Figure 5.3. Adjustment of x and z shims. The spectra show the shape of a single line resonance given by a non-spinning sample. The figures quoted are the dial readings of the potentiometers controlling the shim coil currents.

(a) z shim set at 7.5, x shim varied in steps of 0.5 units; the optimum setting of the x shim is about 4.0.
(b) x shim set at 4.0, z shim varied in steps of 0.5 units; optimum setting of z is about 6.5.
(c) With z shim set at 6.5, x was varied between 4.5 and 3.5 in steps of 0.1 units, the best setting was found with x at 4.3. The next stage would be to spin the sample and to adjust the y shim and the curvature control.

The sweep rate used to record the spectra in (a) to (c) was too slow for any ringing to be displayed. When the peak was observed on the oscilloscope (using a rapid sweep) three or four beats were detectable with the settings used in (c)

that only the y gradient and curvature controls need frequent alteration, although it is best to check also the x and the z gradient settings each morning. These last two are adjusted without the sample spinning, and the criterion adopted for setting them is that a selected sharp line in the spectrum of the standard sample should have maximum height with optimum 'ringing' pattern. This pattern is the exponential decay envelope that follows a signal that is traversed rapidly, and is conveniently observed on the cathode ray oscilloscope. Some typical patterns given by a single line from a non-spinning sample are shown in Figure 5.3 for various setting of the x and z shim controls. It is important to remember that these two controls will probably interact to some extent, so that the combined optimum setting is obtained by repeated alternate adjustments.

Once the x and z shim controls have been optimised the spinner is switched on, and the y control is adjusted for the best ring pattern on the oscilloscope. The appearance of this is noted mentally and the setting of curvature control is altered. The y setting is re-optimised and the two sets of patterns are compared. This procedure is then continued until both the curvature and y gradient controls are optimised simultaneously. It may be necessary to return to the adjustment of the x and z controls, but often this will not be so. Figure 5.4 shows ringing patterns given by the CH_3 resonance in acetaldehyde for different settings of the y and curvature controls, with the x and z controls optimised. The signal is actually a narrow doublet and the beats arise as a result of interference between the two decay envelopes. The pattern to be aimed at is one in which the overall decay envelope appears to be exponential, while the number of beats is maximised.

On spectrometers with field-frequency locking systems the amplitude of the signal used to actuate the locking circuit is a maximum when the resolution is optimised, and this offers a convenient means of monitoring resolution. However, it is important to observe the ringing pattern occasionally in order to avoid the condition of a 'split' magnetic field. In this condition there are quite large magnetic field gradients in the region of the sample, and each line in the spectrum is split into several components which may be quite sharp. A measurement of the strength of the locking signal may give no indication that this has occurred, while the appearance of the ringing pattern will show it up immediately. Normally the locking signal will be a single peak giving an exponential decay pattern, whereas if the field is split a beat pattern will be given.

Once satisfactory resolution has been obtained the standard sample can be replaced by the experimental one, and a final adjustment made to the y gradient shim control, either by observing the ringing pattern of a suitable line, or by monitoring the locking signal.

Recording the spectrum

The following adjustments will generally vary from one sample to another, and are particularly critical for high resolution proton and ^{19}F spectra.

Level of B_1: *i.e. radio-frequency amplitude.* Normally this must be low enough to avoid saturation (p. 7). The gross observable effects of saturation

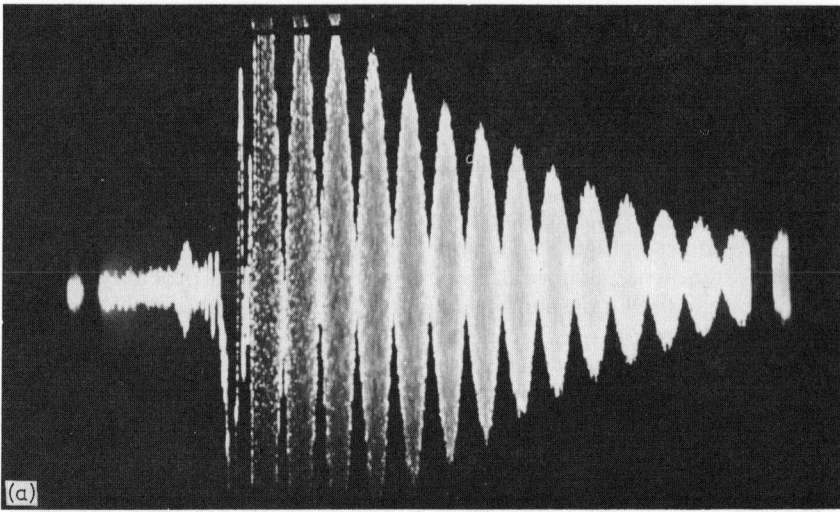

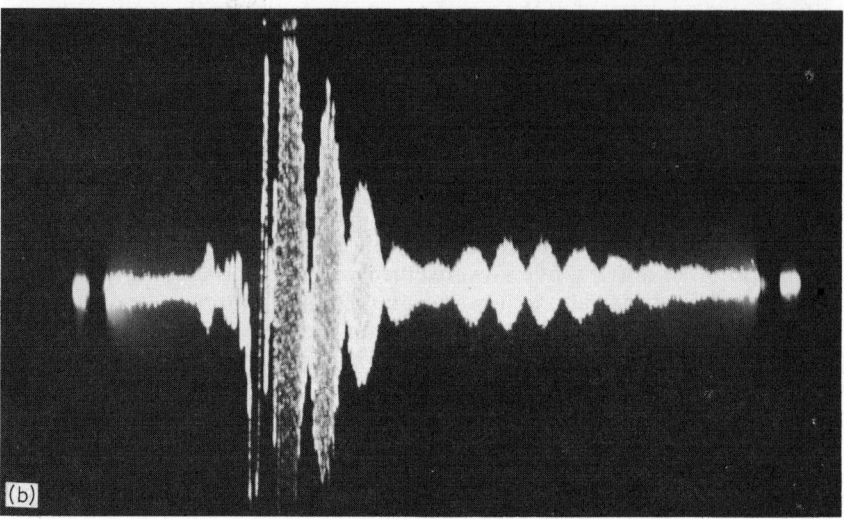

are a general broadening of the spectrum with associated loss of peak height and resolution, as shown in Figure 5.5. A more subtle effect is that the extent of saturation depends *inter alia* upon the transition probability of the observed line, and the relative intensities in tightly coupled spin systems can be seriously perturbed by quite small degrees of saturation. The degree of saturation depends upon the level of B_1, and also on the rate at which the line is traversed. The latter will depend upon recorder sweep-time and sweep-width, and alteration of either of these can be used to assess whether significant saturation is occurring.

When saturation is negligible a halving of the sweep rate should have an inappreciable effect upon peak height. For dilute samples in which it may be difficult to attain a satisfactory signal-to-noise ratio it is often advantageous

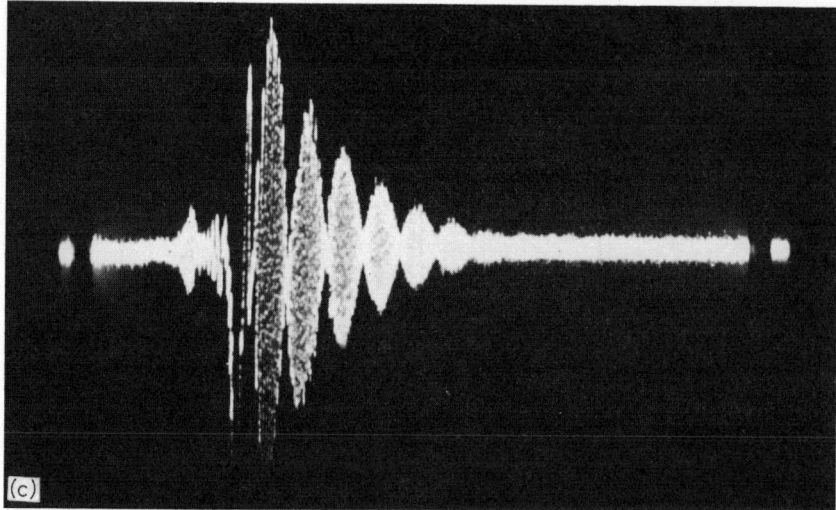

Figure 5.4. Beat patterns. Photographs of oscilloscope traces show the beat patterns produced when the methyl proton doublet of acetaldehyde is scanned rapidly. (a) Optimum settings of *y* and curvature shims. The beat pattern is so extended that the exponential decay was not complete before the oscilloscope scan terminated. (b) The curvature shim setting is optimised, while the *y* control is set too high. After a rapid exponential decay the beat pattern increases in intensity and then decays again. (c) The curvature shim is off-set, while the *y* shim is optimised. Although there is a smooth exponential decay of the beat pattern, the amplitude drops off more rapidly than in (a)

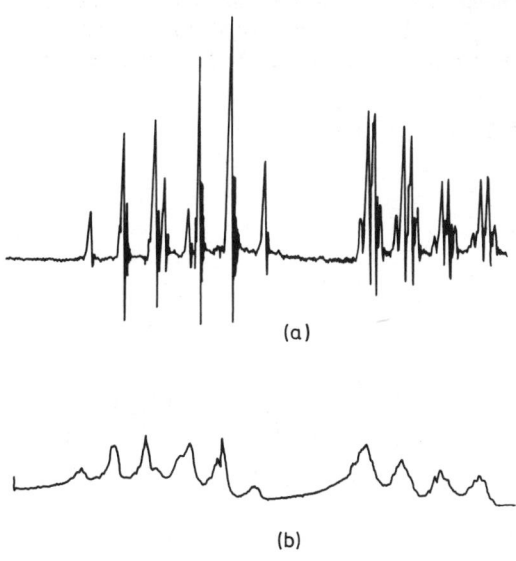

$B_0 \longrightarrow$

Figure 5.5. Effects of saturation. (a) Spectrum run under normal conditions using a sufficiently low r.f. power to avoid saturation. (b) Spectrum obtained when the r.f. level is high enough to cause saturation. The peaks are broadened, causing degradation of the signal-to-noise ratio and loss of resolution

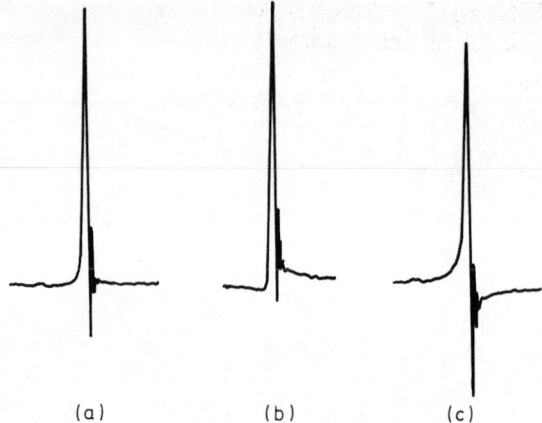

Figure 5.6. Phase adjustment: (a) correctly phased signal; (b) and
(c) incorrectly phased signals

to use a rather high radio-frequency amplitude together with a fairly high sweep rate to minimise saturation. For samples in which the longitudinal relaxation time, T_1, is relatively short, perhaps owing to the presence of paramagnetic ions, higher levels of B_1 may be used, and this can be useful for attaining good signal-to-noise ratio in studies of certain transition metal complexes.

Phase adjustment. This determines the shape of the recorded signal and depends on both the balance of the probe and the setting of the *phase shift control* of the transmitter–receiver unit. The latter essentially controls the

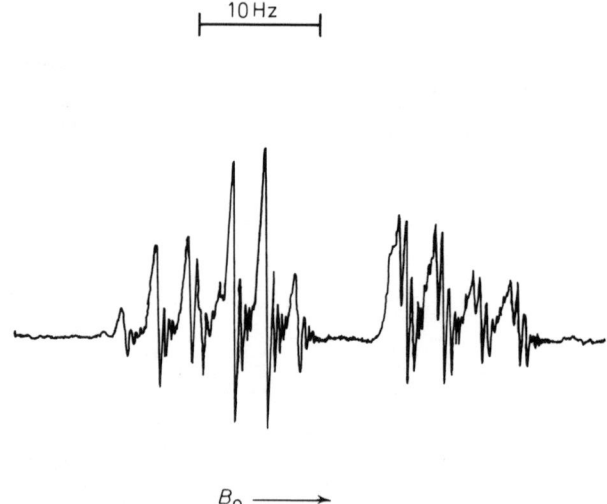

Figure 5.7. Effect of sweep rate upon the appearance of an n.m.r. spectrum. This is the same spectrum as shown in Figure 5.5 (a) except that, owing to too rapid a sweep rate, the peaks appear to be broad-ened, fine structure is lost, and the ringing pattern is accentuated

relationship between the two .signals fed to a phase-sensitive detector. Figure 5.6 shows examples of badly and correctly phased signals, and it is as well to remember that the phase may be sensitive to the level of B_1, so the final adjustment should be done after the B_1 level has been selected. Correct phase adjustment is specially important when the spectrum is to be integrated electronically. When spectra recorded in the frequency-sweep mode extend over a wide range it may be found that the proper phase adjustment varies from one end of the spectrum to the other, and at least one spectrometer manufacturer offers a device to correct for this automatically.

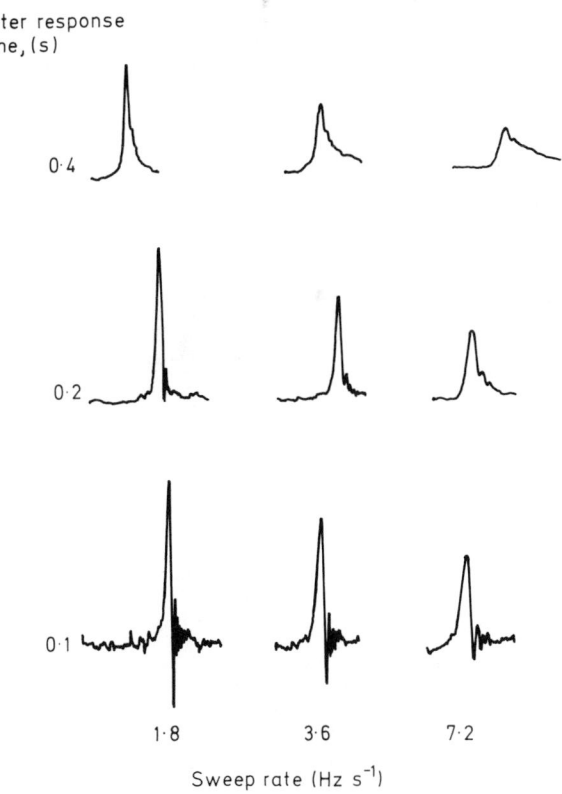

Figure 5.8. Effect of sweep rate and response time on the appearance of a spectrum

Sweep rate and time constant. These are of great importance in determining the appearance of the spectrum and, in particular, spectra which are recorded with too rapid a sweep rate will suffer from loss of fine structure and accentuated ringing patterns as shown in Figure 5.7. For weak samples where problems of signal-to-noise ratio arise, it is desirable to use a time constant of about 0.5 s in the recorder output circuit to cut down the noise. However, it will then be necessary to use an even slower sweep rate to avoid distorting the line shape. With very weak signals one aims at a balance between a small degree of saturation, a moderately slow sweep rate, and a fairly long time constant to cut down the noise. Figure 5.8 shows the effect of using different time constants and sweep rates to record the same line. The exact position at which a line appears on the chart paper depends on the sweep rate, so it is important that the entire spectrum should be recorded at the same speed (Figure 5.9).

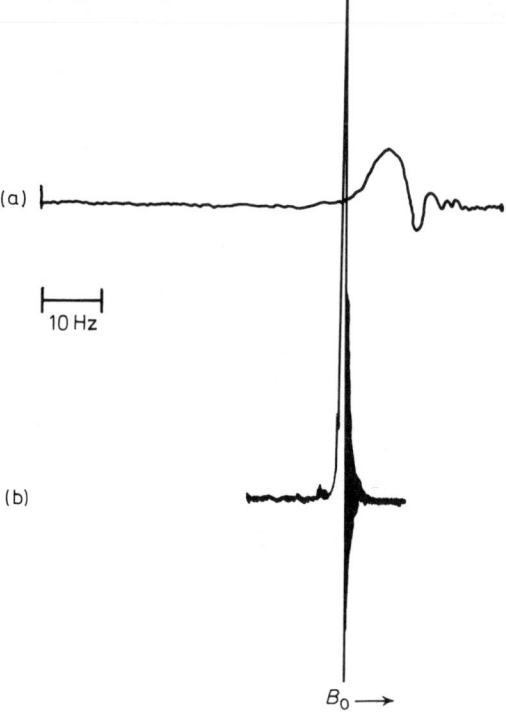

Figure 5.9. Effect of sweep rate on the position of lines in the recorded spectrum. The spectra show the effect of scanning a single line (a) with a very rapid sweep and (b) with a slow sweep. At the higher sweep rate the line appears to be broadened and to have moved to high field, i.e. in the direction of scan

A simple way of improving signal-to-noise ratio is to use a level of B_1 below that which gives any saturation for the particular sample together with an extremely slow sweep rate (say 0.02 Hz s^{-1}) and a correspondingly long recorder time constant. This is known as the SNAIL technique, and can give results comparable to those obtained from a computer of average transients (c.a.t., see p. 79) although it places considerable demands on environmental and instrumental stability.

Spinning rate

If this is too low rather strong sidebands of each line in the spectrum will be produced. These are known as *spinning sidebands*, and their separation from the parent line is equal to the spinning rate. They may be identified by altering the spinning rate when their position will be found to change. Their intensity diminishes as the spinning rate is increased, but too high a rate may lead to vortexing of the sample and loss of resolution, and a satisfactory compromise is about 2500 rev./min. The height of spinning sidebands also depends on shim coil adjustment—particularly the x and z gradient controls—and sample tube quality; under good conditions they should be substantially less than 1% of their parent peak.

5.3 Referencing

Resonance positions in n.m.r. spectra are reported relative to a suitable reference compound. This reference may be either dissolved in the sample (internal reference), or contained separately in a small tube which is concentric with the main sample (external reference), as indicated in Figure 5.10. In the case of an external reference it is necessary to apply a correction for the difference in bulk magnetic susceptibilities of the reference and the sample. This correction can be calculated from Eq. (5.1), which is also the basis of a method for the determination of magnetic susceptibility* by n.m.r. (p. 79).

$$\delta_{\text{corr.}} = \delta_{\text{obs.}} + \frac{2\pi}{3}\left(\kappa_r - \kappa_s\right) \tag{5.1}$$

where κ_s and κ_r are the bulk magnetic susceptibilities of the sample and reference respectively, and the factor $2\pi/3$ arises because this equation applies to long concentric cylindrical samples. For concentric spherical samples the correction is zero, as it also is when an internal reference is used. It is important to realise that if the external reference is mixed with some other material (e.g. a dilute solution of tetramethylsilane in carbon tetrachloride) then the bulk susceptibility of the mixture must be used for κ_r. Table 5.3 gives the bulk magnetic susceptibilities of a number of commonly used n.m.r. solvents. It is usual to ignore the susceptibility correction for nuclei other than the proton, because its magnitude is just the same (typically 1 p.p.m. or less) while the chemical shift differences are much larger.

In proton work the need to apply the foregoing correction may be avoided conveniently by using an internal standard, and the most popular choice is tetramethylsilane (T.M.S.), $(CH_3)_4Si$. The advantages of this material as a reference are: the possession of a large number of magnetically equivalent protons in the molecule, leading to a single intense sharp line; a high degree of magnetic isotropy in most solvents; a proton chemical shift which is outside the range found for most organic compounds so that overlapping of signals is avoided; it is non-polar; unreactive; miscible with or soluble in a wide range of organic solvents; it has high volatility leading to ease of removal from samples. This last feature is sometimes inconvenient, and it is generally advisable to store T.M.S. (and a dropping pipette) in a refrigerator. Normally, the T.M.S. is added to the sample after it has been placed in the n.m.r. tube and one or two drops (to give a total concentration of about

---

* Since existing data on magnetic susceptibilities are in c.g.s. units, equations expressed in the form appropriate for c.g.s. units will be used wherever susceptibilities are involved.

1%) should be sufficient, although if it is intended to use the T.M.S. signal to actuate field-frequency locking circuits it may be advantageous to add rather more. T.M.S. is insoluble in water and for aqueous solutions the sodium salt of 3-(trimethylsilyl)-propane sulphonic acid ($Me_3SiCH_2\text{-}CH_2\text{-}CH_2\text{-}SO_3Na$), which has a methyl proton chemical shift of only 0.02 p.p.m. to low field of T.M.S. itself, has been proposed as an alternative. The other protons in the anion give weak resonances which seldom cause significant

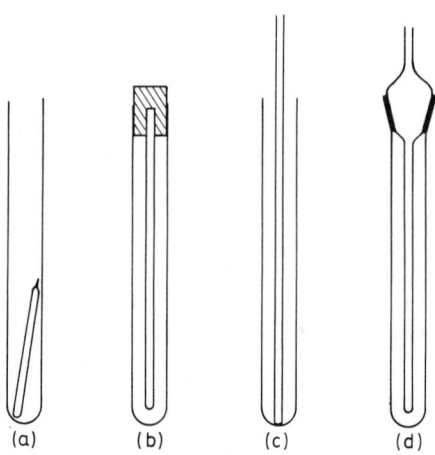

Figure 5.10.   N.m.r. tubes for use with an external reference compound.

(a) The simplest method is to drop a small sealed capillary tube containing the reference compound into the n.m.r. sample tube. Although this is generally satisfactory, it may be found that spinning sidebands are intolerably large. Susceptibility effects depend on the geometry of the cell; since it is impossible to ensure that the reference tube stays in a fixed position relative to the sample tube, this arrangement is unsatisfactory if precise measurements are to be made.

(b) The capillary reference tube is held in a slit cut in a rubber plug that fits tightly into the n.m.r. sample tube. The rubber plug may be cut from a bung by a No. 2 cork borer. If, on spinning, the lower end of the capillary whips excessively it may be helpful to use a small nylon or P.T.F.E. support at the bottom of the sample tube, to hold the end of the reference capillary.

(c) The reference capillary is sealed to the base of the n.m.r. sample tube, if necessary the capillary is also supported by the cap used to close the sample tube. Although this arrangement ensures a constant geometry for the cell, it is not convenient in use, e.g. cleaning the annular space is particularly difficult.

(d) The top of the n.m.r. tube is flared out to form the socket of a small ground glass joint. The capillary is sealed to a small cone made to match the socket. The two parts are ground together (using carborundum powder) to give a snug fit

interference, unless the sample concentration is low and high spectrometer gain is used. The sodium salt of trimethylsilyl tetradeuteriopropionic acid (D.S.S.) has also been suggested as a reference for aqueous solution.

The spectra of other nuclei are often calibrated with respect to external references, and these are dealt with in Chapter 7. However, it is worth pointing out that by means of heteronuclear double resonance experiments the resonances of many nuclei can be referred to an *internal* T.M.S. standard in a wide range of compounds.

It is now necessary to discuss more fully the way in which chemical shifts may be expressed in terms of the field independent quantity δ introduced in Section 2.3. If a sample S gives an n.m.r. signal at a field B_s, and the internal reference compound R gives a peak at B_r then the *magnitude* of the chemical shift is given by

$$\left| \frac{B_r - B_s}{B_r} \right| \times 10^6$$

There are two sources of confusion:

(a) what compound is to be used as the reference compound?
(b) what sign is to be associated with δ?

To a considerable extent (a) has been settled as far as proton resonances are concerned, although it should be remembered that, in the earlier literature, several different reference compounds were used. For other nuclei there is greater latitude in the choice of reference compounds.

Turning to (b), as yet no firm conclusion has been reached about the convention to be adopted for the sign of δ. Earlier workers used the two possible conventions indiscriminately, and care is necessary when dealing with data obtained from older papers. The two possible conventions are:

(i) that δ is defined as follows

$$\delta = \frac{B_r - B_s}{B_r} \times 10^6 \qquad (5.2)$$

so that δ is +ve for peaks shifted to *low field* of the reference peak; or in a frequency-sweep experiment δ is +ve for peaks shifted to *high frequency* of the reference.

(ii) the definition of δ is

$$\delta = \frac{B_s - B_r}{B_r} \times 10^6 \qquad (5.3)$$

so that δ is +ve for peaks shifted to high field (low frequency) of the reference peak.

More coherent opinion existed for a time when the τ scale was introduced. In this, the shift of T.M.S. is given an arbitrary value of 10; for a peak at lower field than that of T.M.S. the τ value was obtained by subtracting $|\delta|$ from 10, while a peak at high field to T.M.S. would have a τ value of $10 + |\delta|$. Subsequently, manufacturers included on their chart paper scales marked in δ the T.M.S. peak having $\delta = 0$. An increasing number of n.m.r. spectroscopists expressed a preference for convention (i), which has certain advantages when used in the mathematical analysis of spectra. At present an authorative decision on the convention to be used is awaited.

The situation is equally chaotic where nuclei other than protons are concerned, and in any particular case it is important to check the convention used by an author. Shifts quoted here will be in terms of δ with shifts to low field of T.M.S. ($\delta = 0$) being positive. Where appropriate τ values will be included in brackets after the δ value.

5.4   Calibration of spectra

Modern spectrometers often use pre-calibrated charts, but these are not utterly reliable and each user must determine for himself how often it is necessary to perform calibration checks. For routine work on a reliable instrument once a day should suffice, but for precise measurements (say in detailed solvent studies) each spectrum must be individually calibrated. A simple routine method of checking the calibration accuracy of a spectrometer is to record the spectrum of a standard sample giving a series of resonances at known positions. A suitable mixture might be: $CHCl_3$ (7.3), CH_2Cl_2 (5.3) Dioxan (3.6) *cyclo*-hexane (1.4), and T.M.S. (0). Approximate δ values are given in parentheses, but the spectrum of the actual mixture used would have to be calibrated initially by one of the precise methods outlined below.

Accurate calibration may be accomplished by using a precision audio-frequency oscillator to modulate either the output of the r.f. transmitter or the magnetic field. This generates a series of sidebands spaced at the modulating frequency of *all* the lines in the spectrum, as shown in Figure 5.11.

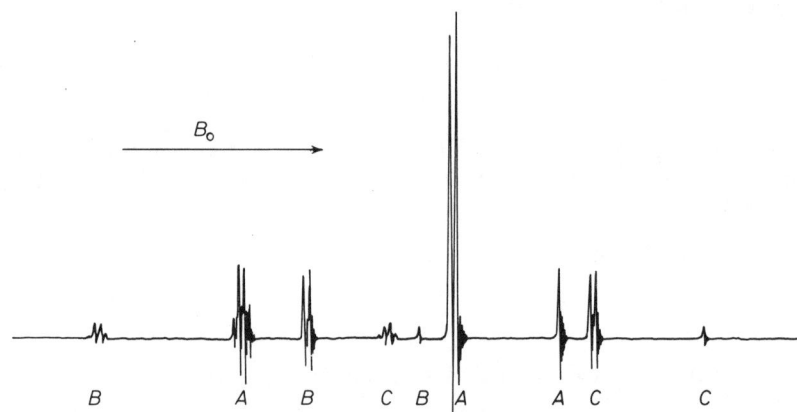

Figure 5.11. ^{1}H n.m.r. spectrum of CH_3CHCl_2 containing a small amount of tetramethylsilane, showing sidebands produced by modulation with an audio frequency of 160 Hz. Peaks labelled *A* form the centreband spectrum, while those labelled *B* and *C* are the lower and upper sidebands respectively. The approximate line positions may be determined from this spectrum using the method shown in Figure 4.4

Clearly this can lead to some confusion in a complicated spectrum, so care must be exercised in the choice of the audio frequency. The total signal intensity is distributed among the original centreband and the sidebands, so the modulating power should be kept as low as is consistent with producing adequately strong sidebands. If very high amplitudes of the audio frequency are used second and higher order sidebands may be obtained. It is usually sufficient to record two or three sidebands throughout the spectrum and obtain line positions by interpolation, as shown in Figure 5.12, but for really precise work the modulating frequency may be varied until the sideband is accurately superposed on the line of interest. This is illustrated in Figure 5.13. It is unwise to alter the modulating amplitude during a calibration

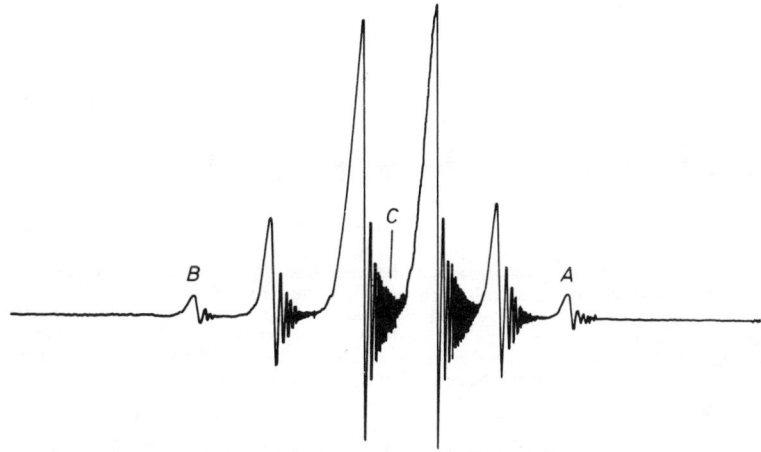

Figure 5.12. ^{1}H n.m.r. spectrum of the $CHCl_2$ group in CH_3CHCl_2, with 340 (A) and 370 (B) Hz sidebands from tetramethylsilane. The distance between A and B is equivalent to 30 Hz. If C is the centre of the multiplet, then C is

$$\frac{AC}{AB} \times 30 = 14.1 \text{ Hz}$$

from the sideband A, and 354.1 Hz from the signal from tetramethylsilane. This method enables the peak positions to be determined more accurately than in Figure 5.11

sweep as this may produce an overall shift in the line positions. The sidebands produced by modulation are not separated from the centreband by exactly the modulating frequency, but their positions also depend upon the amplitude of the r.f. field. However, at power levels commonly encountered any necessary correction is negligible. The calibration of spectra obtained on instruments with internal field-frequency locking is particularly simple and will be described later (p. 68).

5.5 Integration

The measurement of the areas under recorded peaks can be done manually with the aid of a planimeter, or by carefully cutting out the peaks and weighing the paper. However, it is much more convenient to use the electronic integrator with which most modern instruments are equipped. Operation is simple, but three preliminary adjustments are important:

(a) The normal spectrum is recorded and a check is made to ensure that saturation is negligible (p. 55)
(b) The phase is very carefully adjusted to give a pure absorption signal for the normal spectrum

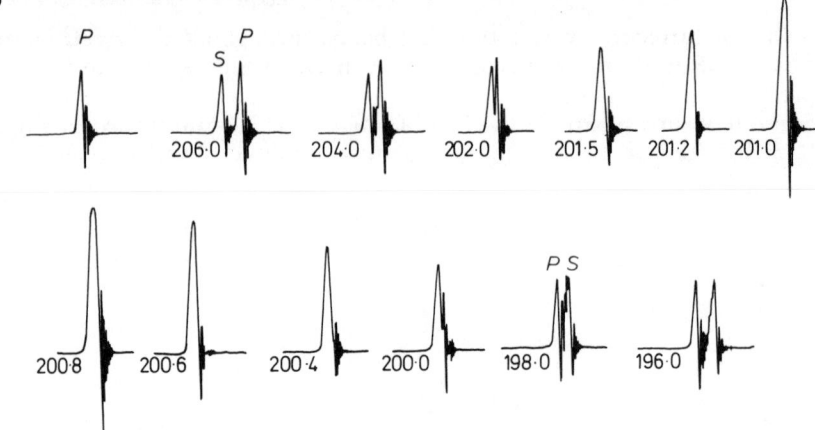

Figure 5.13. Determination of the separation between lines by the method of sideband super-position. The first spectrum shows the single line, P, in the ^{1}H spectrum of methylene chloride. Subsequent spectra show the peak P and S, a sideband from tetramethylsilane, at different modulation frequencies. By adjusting the modulation frequency it is possible to superimpose the sideband on peak P. This experiment gives a separation of 200.8 Hz as the separation between the peak P and tetramethylsilane

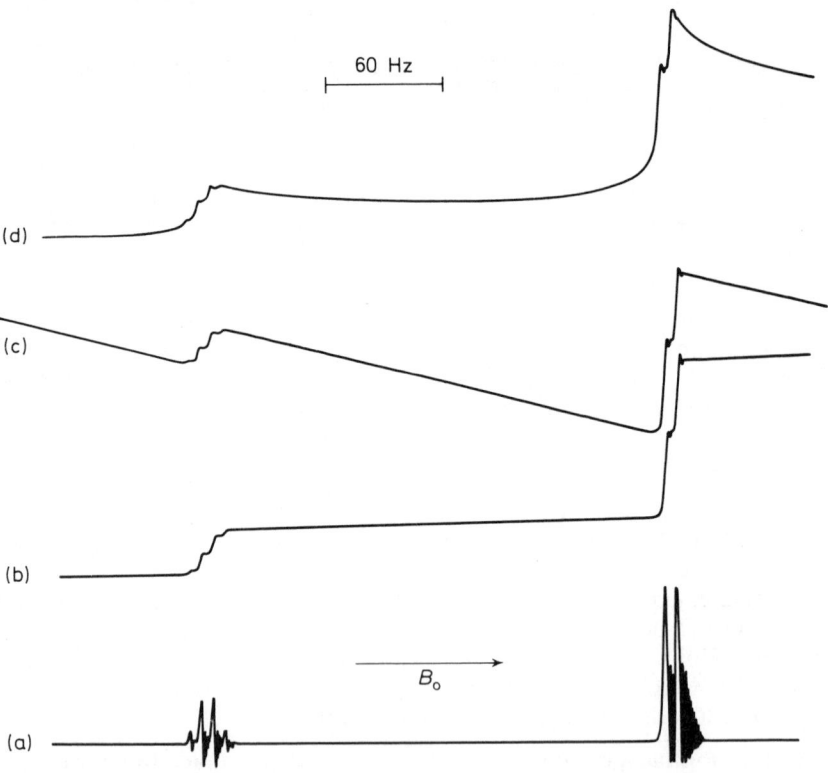

Figure 5.14. Integration. (a) The absorption spectrum to be integrated. (b) Integral trace obtained when phase and drift controls are correctly set. (c) Integral trace obtained when the phase is correctly set but the drift is not properly balanced. (d) Integral trace obtained when the drift is correctly balanced but the phase is not set properly

(c) The recorder pen is set clear of any line in the spectrum, the integrator is switched on, and the gain is set fairly high.

The *drift* or *bias* control is then adjusted until there is no pen movement, the gain is reduced to a suitable level, and the integrated spectrum is recorded at a fairly high sweep rate to minimise residual drift and saturation. It is advisable to record the integral several times (cancelling the accumulated signal between each successive run) and to take an average. It is important to switch off the integrator after use otherwise a large signal due to random noise will accumulate. The integral trace is obtained as a step function as shown in Figure 5.14, in which the relative areas are given by the heights of the appropriate steps. This figure also shows the effects of incorrect drift and phase adjustment. The heights of the steps depend also upon the sweep rate, so this must remain constant throughout a particular run.

5.6 Spectrometers with internal field-frequency locking systems

The foregoing discussion has dealt with techniques and adjustments that are applicable to all types of spectrometer, including those with external field stabilisation systems. The highest degree of stability, however, can only be attained by *locking* the field with the aid of a signal derived from the experimental sample itself, and we now consider some additional points in connection with such systems.

T.M.S. (1–10%) added to the sample for reference purposes is often a convenient source of the locking signal, but other substances, including the solvent itself may be suitable. The main characteristics demanded of the locking signal are: it must be a single sharp line; it must not be close to another strong line in the spectrum, otherwise the field may lock on to this signal in preference to the desired one; it should be reasonably distant (more than say 60 Hz) from line of interest in the spectrum irrespective of whether they are strong or weak.

Once the locking signal has been chosen the normal procedure is to display the selected line on the cathode ray oscilloscope, and trim up the resolution in the usual way (p. 63). The phase is then adjusted to give a derivative line shape as in Figure 5.15 (a) or (b), according to the design of the spectrometer. This is necessary to enable the locking circuits to distinguish between drifts of the magnetic field to high or to low field. Finally, the repetitive sawtooth sweep is switched off, the field is adjusted slightly until a movement of the spot on the cathode ray oscilloscope indicates exact resonance of the selected line, and the locking circuit is switched on. The y gradient and curvature homogeneity controls may now be finally optimised by monitoring the height of the locking signal either on the cathode ray oscillosope or on a separate meter. There are three common causes of failure to obtain a satisfactory lock:

(a) A rapid oscillation of the amplitude of the monitored locking signal indicates that the gain of the audio phase-sensitive detector amplifier is set too high

(b) A slow pulsation of the locking signal can be due to incipient saturation arising from too high r.f. amplitude

(c) Imperfections in the sample tube or too high a spinning rate can affect the line shape and cause locking instability, even though the signal may look normal when scanned on the oscilloscope.

The adjustments of the second audio-frequency oscillator and amplifier used for recording the spectrum follow those outlined previously for simpler spectrometers. A frequency-sweep spectrum is obtained by changing the frequency of the oscillator used for recording as the pen traverses the chart, and a field-sweep spectrum by changing the frequency of the oscillator used to excite the locking resonance. The latter method of operation is somewhat

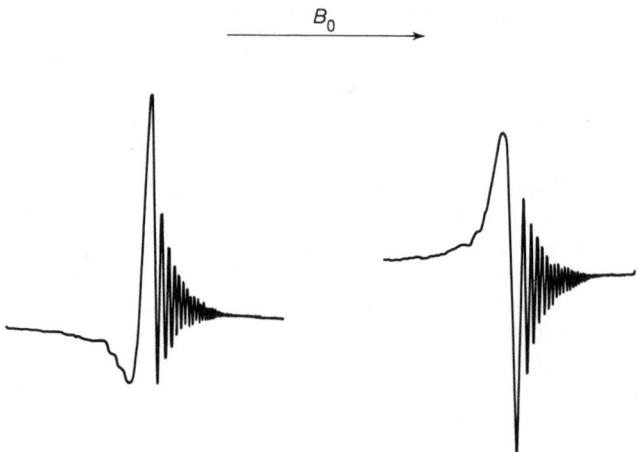

Figure 5.15. Derivative line shapes for the locking signal when using an internal field-frequency lock. The design of the electronic circuits forming the locking system will determine which of the alternative shapes should be used on a particular instrument

less stable as too high a sweep rate may lead to loss of lock. This kind of system gives highly reproducible results, and there is normally no difficulty in monitoring the height of any selected line in the spectrum for indefinite periods provided saturation is avoided. This is often known as 'sitting on' a line and is useful for certain types of double resonance experiment (p. 95).

Calibration is achieved simply by stopping the pen at any desired point and measuring the frequency of the variable frequency oscillator to 0.1 Hz with a suitable frequency counter. This applies irrespective of whether the field— or the frequency—sweep mode of operation is used, and line positions may be established either by interpolation or by direct measurement of the line position. If desired the simpler sideband method described earlier may be used for calibration. In connection with spectra recorded in frequency-sweep mode it is important to remember that highly shielded nuclei (i.e. those giving a resonance at *high field*) will give signals at a *lower* radio frequency. However, this may correspond to a higher or lower audio modulation frequency according to whether the low- or high-field modulation sidebands are used for observation. As the resonance of the compound used to provide the locking signal is traversed the two audio frequencies will coincide and a characteristic 'beat' pattern will be produced as shown, for example, in Figure 5.16. It is unwise to allow the pen to dwell long in this region otherwise the lock may be lost.

Integration of spectra recorded on instruments with internal field-frequency locking systems is done in the usual way, but more accurate results are obtained if the spectra are field-swept ones. This is because the change in frequency of the recording audio oscillator in the frequency-sweep mode is accompanied by a small change in the amplitude of the actual sideband used to record the signal and so the intensity is affected. The problem is less serious for instruments which use relatively high frequency (~ 50 kHz) modulation sidebands.

5.7 Measurements at other temperatures

This facility, now available on most commercial instruments, is very important for kinetic and conformational studies. In some instances variation of the sample temperature may improve the appearance of the spectrum. At temperatures above ambient, the viscosity of the sample is reduced, leading to a decrease in line broadening. At higher temperatures the solubility of most materials is increased appreciably, leading to improved signal strength for compounds of low solubility. In the case of solids that are insoluble in acceptable solvents, it may be possible to obtain spectra from the molten sample. Low temperatures are mainly of importance in detailed kinetic studies; however, cooling the sample may decrease the rate of an exchange reaction and lead to an improvement in the line-widths in the spectrum. In addition to intensity differences due to changes in solubility with temperature, differences will occur as a result of temperature perturbation of the Boltzmann distribution of the nuclear spins.

At low temperatures, the population of the lower energy state will be increased (relative to the population at room temperature) leading to increased signal strength. At high temperatures, the population of the upper energy state is increased (the populations of upper and lower energy states become more nearly equal) and the signal intensity is decreased. If the only factor in influencing signal intensity is the Boltzmann distribution, then the signal intensity is approximately proportional to the reciprocal of the absolute temperature. Changing the temperature of a suitable sample from room temperature to $-123\,°C$ will approximately double the intensity; changing the sample temperature from $-40\,°C$ to $+200\,°C$ will result in the signal intensity being approximately halved.

Variable temperature systems have been described briefly in Chapter 4. In practice, the sample temperature may differ slightly from that recorded by the thermocouple monitoring the gas stream. For accurate work it may be necessary to calibrate the system either by a thermistor probe or a second thermocouple placed in the spinning sample; secondary calibration may be achieved by measuring a known temperature-dependent chemical shift. The separation between the two peaks in methanol can be used as a standard for low-temperature calibration, while ethylene glycol is suitable for high-temperature work. A sealed capillary tube containing the standard liquid can be placed in the sample tube; the recorded spectrum will include peaks from the standard, and measurement of the separation between appropriate peaks enables the temperature of the sample to be estimated. The calibration of methanol and glycol n.m.r. thermometers has been discussed by Geet[1]. The chemical shift Δv (in Hz) at 60 MHz between the CH_3 and the OH peaks over the range 220–330 K is given (with r.m.s. error 0.6 K) by

$$T = 435.5 - 1.193 \,|\Delta v| - 29.3\,(10^{-2}\,\Delta v)^2 \tag{5.4}$$

Over a temperature range of 60 K the following linear relationships represent the data with errors less than 0.8 K:

$$\text{Range} \quad 220\text{–}280 \text{ K} \qquad T = 478.6 - 1.906\,|\Delta v| \quad \begin{bmatrix} \text{Error} \\ 0.8\text{ K} \end{bmatrix} \tag{5.5 a}$$

$$260\text{–}320 \text{ K} \qquad T = 464.0 - 1.775\,|\Delta v| \quad \begin{bmatrix} \text{Error} \\ 0.4\text{ K} \end{bmatrix} \tag{5.5 b}$$

Duerst and Merbach[2] have suggested the use of a mixture of 51.8% water, 48.1% methanol, and 0.1% hydrogen chloride. They found that between -40 and 80 °C the temperature t °C is related to S the shift in parts per million between the two resonance peaks by:

$$t = 160.00 - 90.50S \tag{5.6}$$

The temperature coefficient depends on the composition of the sample. Geet finds for temperatures between 310 and 410 K the chemical shift Δv between the CH_2 and OH peaks of the ethylene glycol is given by:

$$T = 466.0 - 1.694\,|\Delta v| \tag{5.7}$$

with an error of 0.3 K.

A similar equation

$$T = 466.5 - 1.691\,|\Delta v| \tag{5.8}$$

has been suggested by Neuman and Jonas[3].

The techniques of measurement are the same as for work at ordinary temperatures although the homogeneity—notably the y gradient—and probe balance may need readjustment. The latter can be avoided to some extent by using a lower r.f. amplifier gain setting than usual. It is important to allow the system to come fairly close to thermal equilibrium before attempting to make measurements. Ordinary sample tubes can be used, but if they are to be heated the entire sample should be tested away from the spectrometer by maintaining it for some time a little above the expected maximum temperature. In particular, T.M.S. may generate considerable pressures in sealed tubes at elevated temperatures. It is advisable, where possible, to use plastic caps for closing sample tubes that are to be heated; in the event of a high pressure being generated the cap will normally be blown off before the tube itself is shattered.

At very low temperatures it may be found that sample spinning becomes difficult because of the condensation of moisture from the air which drives the spinner; this can be avoided by using dried air or nitrogen instead. The foregoing techniques will be found suitable over the range -150 to $+200\,°C$; for temperatures outside this range specially constructed probes have to be used. Solvents suitable for variable temperature work are listed in Table 5.2.

Table 5.2 SOLVENTS FOR N.M.R. TEMPERATURE STUDIES

Compound	m.p. (°C)	b.p. (°C)
Solvents for use at low temperatures		
CCl_4	-22	$+77$
$CHCl_3$	-63	$+61$
SO_2	-73	-10
$(CH_3)_2CO$	-95	$+56$
CH_2Cl_2	-97	$+40$
CH_3OH	-98	$+65$
CCl_3F	-111	$+24$
CS_2	-112	$+45$
$(CH_3CH_2)_2O$	-116	$+34$
CCl_2F_2	-158	-30
Solvents for use at high temperatures		
C_7F_8		$+104$
$CHBr_3$	$+8$	$+150$
$o\text{-}C_6H_4Cl_2$	-17	$+179$
CBr_4	$+90$	$+190$
Biphenyl	$+70$	$+256$

5.8 Double resonance experiments[4]

Instrumentally these fall into two categories: homonuclear and hetero-nuclear; and facility for the former is now routinely available on many commercial instruments.

Homonuclear experiments

We shall illustrate the procedure with examples of proton spectra; the extension to other nuclei is straightforward. The simplest kind of experiment to perform is decoupling, in which the fine structure of a particular resonance is removed or reduced. This is achieved by irradiating the transitions of the nucleus responsible for the fine structure with a strong radio-frequency field. In principle, this could be done by using a second variable r.f. oscillator, but this is needlessly expensive, and a simple alternative is to use an additional audio-frequency oscillator to produce a sideband at the required decoupling frequency, v_2, by modulation of the main operating r.f. signal (v) for the instrument. Many commercial instruments currently available use this system, although it does have some disadvantages in the way of producing additional spurious sidebands. The interpretation of double resonance experiments is most straightforward when the spectrum is recorded in the frequency-sweep mode using frequency v_1 for observation, because the decoupling frequency (v_2) then remains at a fixed point relative to the remainder of the spectrum. For instruments equipped with field-frequency locking the general procedure is to record the normal spectrum; the recorder

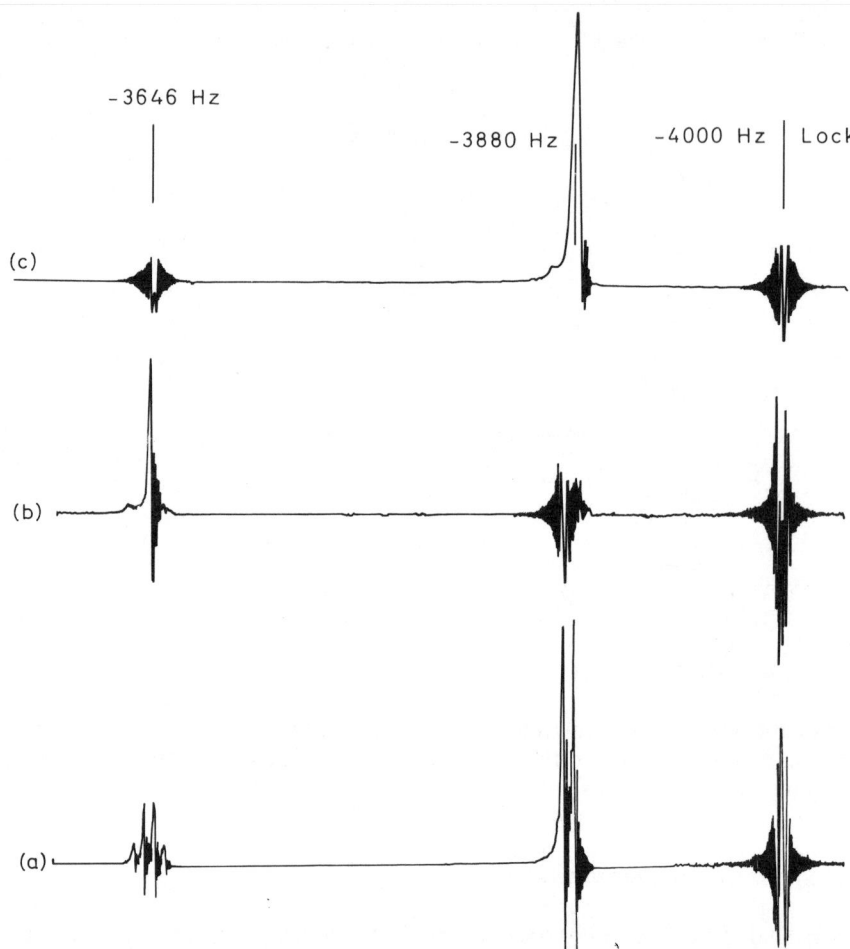

Figure 5.16. Frequency-sweep single and double resonance spectra of 1,1-dichloroethane. A small amount of tetramethylsilane has been added to the sample to provide the locking signal. At the positions of the three markers the sweep sideband frequency was measured with a frequency counter. The negative signs associated with the frequencies indicate that the lower sideband was used in these experiments.

  (a) Single resonance spectrum. Note the beat pattern produced when the locking and sweep frequences are nearly equal.

  (b) Double resonance spectrum with the irradiating field applied at the frequency of the doublet. Note the beat pattern produced between the scan and irradiating frequencies.

  (c) Double resonance spectrum with the irradiating field applied at the frequency of the quartet. The lines of the quartet extend over a greater frequency range than the lines in the doublet; in this case decoupling is less complete, and the line-width of the decoupled CH_3 peak in this spectrum is greater than the line-width of the decoupled $CHCl_2$ proton peak in (b)

pen (and hence the observing frequency) is then returned to the resonance arising from the nucleus to be irradiated by the decoupling frequency. The frequency of the additional audio-frequency oscillator is then adjusted until the pen registers a beat signal between v_2 and the observing frequency (a frequency counter can be very useful here); v_2 is adjusted until the beat frequency is zero at which setting the frequencies v_1 and v_2 are identical. The amplitude of the frequency v_2 is set appropriately, and the double resonance spectrum is scanned by varying the frequency v_1.

Figure 5.16 gives the frequencies of the locking, recording, and decoupling audio oscillators used for double resonance experiments on 1,1-dichloro-ethane, CH_3CHCl_2 containing a small amount of T.M.S. to provide the locking signal. In this case the low-field modulation sidebands were used. For decoupling experiments such as these, quite high amplitudes of B_2 (the field strength of the decoupling frequency) are needed and are obtained by setting the centreband r.f. power level much higher than usual (40 dB = 100 times the amplitude in the present instance) and using high modulating power. The outputs of the audio oscillators used for locking and recording

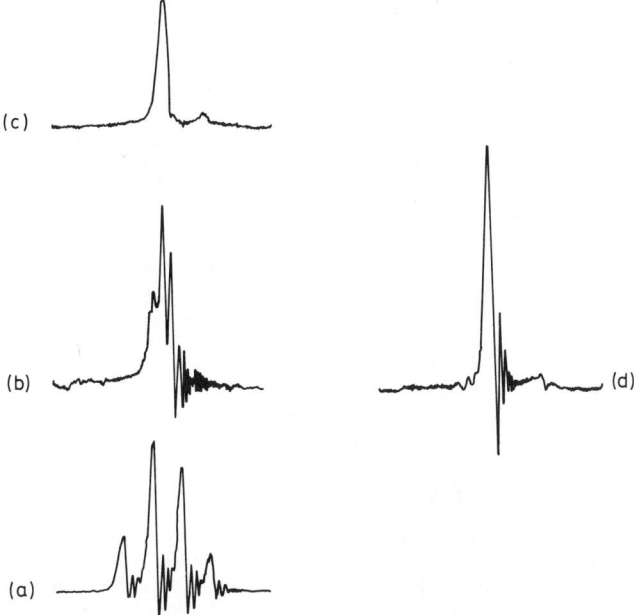

Figure 5.17. The effect of using incorrect frequencies in decoupling experiments. The single resonance spectrum of the quartet of CH_3CHCl_2 is shown in (a). The other spectra show the appearance of this quartet when a decoupling field (of constant power) is applied offset from the CH_3 resonance frequency by (b) 5 Hz; (c) 2 Hz; (d) 0 Hz

are reduced correspondingly to avoid saturation, and it may be helpful also to use lower r.f. gain to minimise difficulties in balancing the probe. The effects of using the incorrect frequency for v_2 and too low an amplitude for decoupling are shown in Figures 5.17 and 5.18 respectively. The decoupling frequency must not approach too closely to the locking signal, and in practice it is not possible to irradiate strongly groups that are closer than 30 Hz to the locking signal. Difficulties may also be encountered from quite strong sidebands at the mains (line) frequency of 50 or 60 Hz.

For spectrometers without field-frequency stabilisation, but which do have baseline stabilisation, the basic procedure is very similar to the foregoing. The spectrum in such spectrometers is often recorded in field-sweep mode, and the difference between the recording and decoupling frequencies must be set equal to the frequency separation of the two groups that are to be decoupled. For example, using a spectrometer with a fixed frequency recording oscillator working at 4000 Hz, irradiation of resonance A, 234 Hz to low field of B, the band to be observed, would require a decoupling modulation

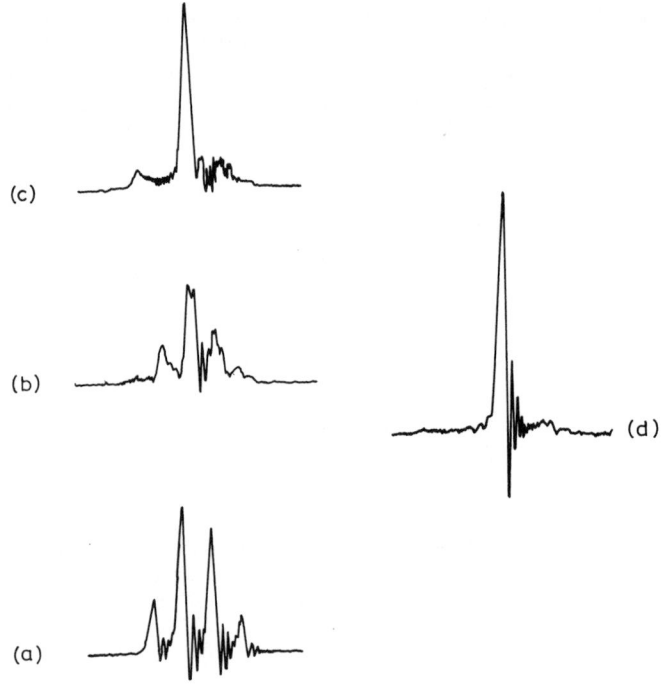

Figure 5.18. The effect of using different power for the irradiating field in double resonance experiments. The single resonance spectrum of the quartet of CH_3CHCl_2 is shown in (a). Subsequent spectra show the appearance of this quartet when a decoupling field is applied at the CH_3 frequency using (b) low power; (c) intermediate power; (d) high power

frequency of 4234 Hz if the low-field sidebands are used. If the roles of resonances A and B are reversed, then 3766 Hz would be the decoupling frequency to irradiate B while observing A. If the high-field sidebands were used in the spectrometer, then frequency 3766 Hz would be used to irradiate A while observing, B, and 4234 Hz would be used for irradiating B while observing A.

In experiments of this type no beat pattern appears since the two modulation frequencies never coincide; nor is it obvious from the recorded spectrum just which group has been irradiated. Of course, if desired, field-sweep double resonance experiments may be performed on spectrometers with internal locking systems, but in general more experiments will be necessary than if frequency sweep is used.

Even the simplest spectrometers without baseline stabilisation may be used for decoupling experiments if an audio-frequency oscillator is available to modulate the magnetic field. This gives a pair of sidebands as discussed in the section on calibration (p. 64), and one of these may be used to record the spectrum while the centreband irradiates the group to be decoupled. The audio oscillator operates at the frequency separation of the groups to be decoupled, the r.f. power is set to a high level and the modulation index (i.e. the relative intensity of the sidebands) is kept low. In a complex spectrum there may be interference from the unused sideband, but this can be avoided by using a single-sideband modulation scheme wherein the unwanted sideband is suppressed.

More sophisticated double resonance experiments, such as selective decoupling and spin tickling, which require weaker r.f. fields may be performed similarly, it generally being convenient to reduce the overall centreband power somewhat and then make fine adjustments to the output of the audio oscillator. Experiments of this type demand greater care in setting the irradiating frequency, and a correspondingly higher stability in the oscillator used to provide this frequency. Their interpretation is considerably simpler if the spectrometer is operated in frequency-sweep mode.

Heteronuclear double resonance experiments

Many proton (and ^{19}F) spectra are complicated by the effects of coupling to another nuclear species with a spin, e.g. ^{14}N or ^{31}P. These effects may be readily removed, and the spectrum thereby simplified, by irradiating at the resonance frequency of the other nucleus, but at present facilities for this are not commonly provided on commercial instruments. Normally, the

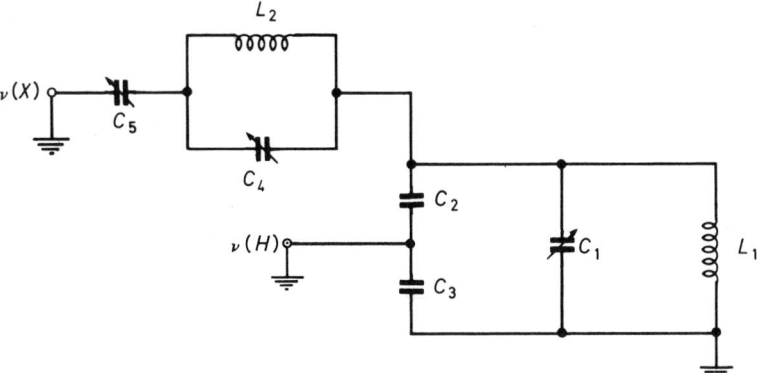

Figure 5.19. Circuit diagram of a double-tuned transmitter coil. The components C_4, C_5, and L_2 are added to the normal transmitter coil circuit. The composite circuit is tuned in the following way: (a) C_1 is adjusted to tune the circuit of L_1 to the frequency $v(H)$; (b) C_4 is adjusted to block the frequency $v(H)$ from the oscillator providing the second frequency $v(X)$; (c) C_5 is adjusted to tune the circuit to the frequency $v(X)$

spectrometer must be modified either by the addition of a second r.f. coil within the probe, or by double tuning the existing transmitter coil. A circuit for the latter is shown in Figure 5.19. The ideal source of the second frequency is a frequency synthesiser (of which several varieties are commercially available) with a power amplifier. However, if a lower degree of frequency stability can be tolerated, or if less versatility is needed, cheaper alternatives may be suitable. For example, an audio-modulated crystal-controlled r.f. transmitter

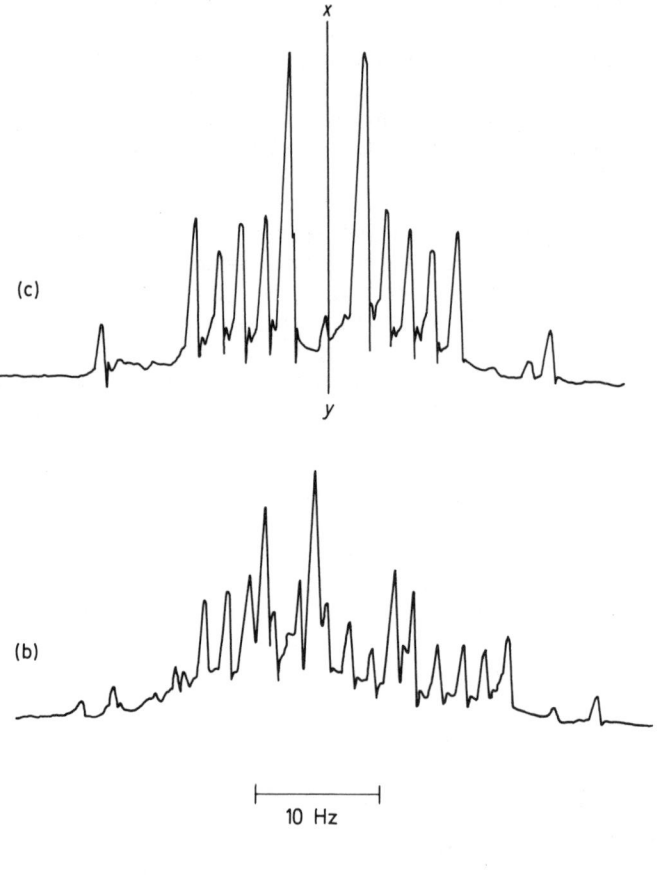

Figure 5.20. 60 MHz spectra of the non-aromatic protons of phenyl-ethylenephosphite (a). (b) Normal spectrum; (c) ^{31}P decoupled to give a spectrum symmetrical about xy which is characteristic of the $AA'BB'$ spin system. The undecoupled spectrum is unsymmetrical because the two coupling constants $J(^{31}P—H_A)$ and $J(^{31}P—H_B)$ are not equal

would suffice for decoupling experiments and chemical shift determination for any chosen nucleus.

The general experimental procedure is to monitor a suitable line in the proton (or ^{19}F) spectrum, and vary the frequency of the decoupling oscillator until a perturbation is observed. Trial spectra are then recorded at slightly different settings of this frequency until a satisfactory result is obtained. An example of ^{31}P decoupling is shown in Figure 5.20. Ideally, the spectrometer should be equipped with a field-frequency locking system to facilitate location of the decoupling frequency, but oscilloscope observation of repeated scans of a suitable line can be satisfactory. If the region of the proton spectrum which is to be recorded is narrow it matters little whether the spectrum is

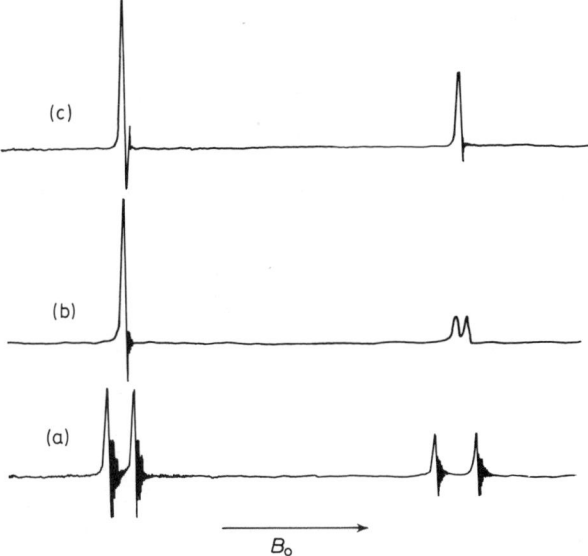

Figure 5.21. Field-sweep proton spectra of methyl dimethyl phosphonate, $CH_3(CH_3O)_2PO$; (a) Normal spectrum; (b) with irradiation at a fixed ^{31}P frequency chosen to optimise decoupling of the methoxy (low-field) protons; (c) with irradiation at a ^{31}P frequency manually adjusted to remain on resonance throughout the experiment

recorded in the field- or the frequency-sweep mode, but for wider scans the latter is better. Alternatively, the decoupling frequency may be readjusted as the proton spectrum is scanned in field-sweep. Equation (5.9) gives the magnitude of the necessary correction, for $^1H-\{X\}^*$ experiments

$$\delta v_X = -\delta v_H \cdot |\gamma(X)|/\gamma(^1H)$$
$$= -\delta v_H \cdot v(X)/v(^1H) \tag{5.9}$$

Where δv_X is the required correction to the decoupling frequency, and δv_H is the change in field strength measured in frequency units *in the proton spectrum*. Figure 5.21 shows field-sweep decoupling experiments in which (b) no correction has been applied, and (c) the decoupling frequency has been adjusted to remain on resonance throughout.

* This symbol indicates that the resonance of ^1H is observed while that of X is irradiated.

Difficulties are encountered when the spectrum of the nucleus to be decoupled extends over a wide range, perhaps because one of the coupling constants is very large, and it may then be impossible to effect complete decoupling by normal means. In these circumstances it is advantageous to simultaneously *modulate* the decoupling r.f., either with random (white) noise or a suitable audio frequency, as this spreads the available power over

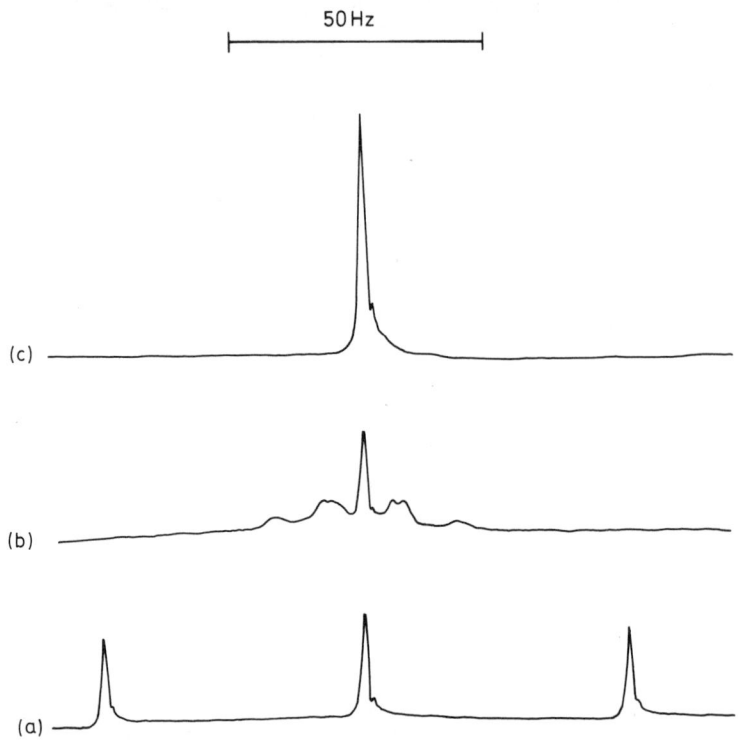

Figure 5.22. The ^{1}H n.m.r. spectrum arising from the ammonium ion. (a) The single resonance spectrum, showing the $1:1:1$ triplet arising from coupling with the ^{14}N nucleus ($I = 1$). (b) Double resonance spectrum, with irradiation at the ^{14}N frequency. The power available in the irradiating field was insufficient for complete decoupling. In addition to the central line there are marked satellites. (c) Using the same power for the decoupling field as in (b) together with audiomodulation, decoupling is essentially complete

a greater range of frequency. As an example of the latter, Figure 5.22 shows ^{14}N decoupling experiments in the ammonium ion NH_4+. The proton spectrum is a $1:1:1$ triplet ($J = 52$ Hz) owing to coupling to ^{14}N ($I = 1$, abundance = 99.6%) and displays no quadrupolar broadening because of the symmetrical environment of the nitrogen atom. A normal ^{14}N decoupling experiment gives a broad central line with marked satellites (b), whereas when the same decoupling power is used together with audio modulation the decoupling is essentially complete.

Decoupling is only one aspect of double resonance, and important results can be obtained by using very low amplitudes of the second irradiating field. The detailed interpretation of such experiments will be dealt with later

(p. 85), and the practical aspects are much the same as for normal decoupling except that the demands on instrumental stability are greater.

5.9 Sensitivity enhancement by spectral accumulation

Although spectral signal-to-noise ratios can be improved by using slow sweep rates together with a long recorder time constant (p. 60), it is generally more satisfactory to use spectral accumulation. The normal spectrum is scanned repeatedly and the individual spectra are added together, the result being stored in the memory of a small computer. Naturally in this process the spectral noise is also accumulated, but since it is random the total noise intensity is proportional to the square root of the number of scans, whereas the total signal intensity is directly proportional to the number of scans. The overall improvement in signal-to-noise ratio is thus proportional to the square root of the number of scans, and an important advantage of the method is that quite high sweep rates may be used since any smoothing may be delayed until the final accumulated spectrum is read out. Furthermore, these higher sweep rates permit the use of fairly high amplitudes of B_1 and this may increase the signal-to-noise ratio by a factor of two over that achieved by a single very slow scan with extensive smoothing.

With sufficiently stable instrumentation (i.e. field-frequency locking and automatic field homogeneity control) it is possible to make overnight runs and so achieve enhancement factors of 20 or more, but for many purposes much smaller numbers of scans are adequate. It is important that all the individual recorded spectra should be accurately superposed, and if the spectrometer does not have a field-frequency locking system this can be achieved by using a signal from the sample to trigger the computer at the beginning of each scan.

Most accumulation devices of this type can also subtract a recorded spectrum, and this feature can be used to compare two very similar spectra. If one is added and the other subtracted then the final record will contain only signals which are not common to both.

5.10 The determination of magnetic susceptibility[5]

As mentioned on p. 61 when an external reference is used the observed chemical shift of a resonance depends on the bulk susceptibilities of the sample and the reference solutions, this can be made the basis of a convenient method for the determination of magnetic susceptibilities. The accuracy of the results obtained by the use of n.m.r. for this purpose can be comparable with that of classical methods. The n.m.r. method has the advantages that:

(a) a smaller sample is required
(b) the measurements are simpler to carry out
(c) the temperature of the sample is easily controlled and can be readily varied.

Details of the method and the calculation of results depend on whether the sample tube is spinning or static. In either case the method requires a coaxial cylindrical tube to be supported within a standard n.m.r. sample tube.

Measurements using a spinning sample

If $\Delta\kappa$ is the difference in volume magnetic susceptibility between two solutions, then Δv the difference in chemical shift (in Hz) of a proton resonance line of an inert solute in the two solutions is given, for a spinning cylindrical sample, by

$$\frac{\Delta v}{v} = \frac{2\pi \, \Delta\kappa}{3} \tag{5.10}$$

in which v is the frequency of the r.f. field.

A convenient reference compound for the study of aqueous solutions is t-butyl alcohol. For investigation of transition metal ions in solution, care must be taken to avoid reference compounds which become involved in specific association with the metal ion.

A dilute solution (about 3%) of the reference compound in a solvent of known magnetic susceptibility is placed in the inner tube; the outer tube contains the solution whose magnetic susceptibility is to be determined, together with sufficient of the reference compound for it to have a 3% concentration. The spectrum given by this composite sample will show two lines from the reference compound, as a result of the difference in susceptibilities of the two solutions. The solution of greater bulk diamagnetic susceptibility will give a reference compound peak at higher field. Table 5.3 compares susceptibilities obtained by the n.m.r. method with those obtained by the Gouy method.

Table 5.3 COMPARISON OF MASS-SUSCEPTIBILITIES χ OBTAINED BY THE N.M.R. METHOD WITH LITERATURE VALUES

Substance	$10^6\chi$ at $20°C \pm 1$ deg	
	n.m.r.[a]	literature
$NiCl_2$ (0.162M)	34.00 ± 0.2	34.20 ± 0.1[b]
$CuSO_4$ (0.0788M)	9.6	10.1[c]
$NH_4[Cr(CNS)_4(NH_3)_2]$		
(0.0152M)	18.6	17.4[d]

[a] Evans, D. J., *Chem. Soc.*, 2003 (1959)
[b] Selwood, P. W., *Magnetochemistry*, Interscience, 25 (1956)
[c] Amiel, J., *Compt. Rend.*, **213**, 240 (1941)
[d] Berkman, S., and Zocker, H., *Z. phys. Chem.*, **124**, 318 (1926)

A difficulty often encountered when a spinning sample is used for susceptibility measurements is the presence of intense spinning sidebands in the spectrum; this problem is avoided in the static method.

Measurements using a static sample

In this method the sample is not spun, and although the results are not as accurate as the classical methods or the spinning n.m.r. method, they are sufficiently good for making susceptibility corrections to chemical shifts obtained with an external reference. A pair of coaxial tubes is again used, and it is generally more convenient to place the reference compound (which must give an n.m.r. signal) in the annular space between the two tubes, while the sample to be measured (which need not give an n.m.r. signal) is placed in the central tube. The reference compound in the annulus gives rise to two n.m.r. peaks, because it experiences a non-uniform magnetic field owing to the different bulk susceptibility of the experimental sample. The separation in Hz between the two reference compound peaks is given by

$$\Delta v = 4\pi v \left[\frac{a^2 \kappa_1}{r^2} - \frac{b^2 \kappa_3}{r^2} - \frac{(a^2 - b^2)}{r^2} \kappa_2 \right] \tag{5.11}$$

where v is the frequency of the r.f. field, κ_1, κ_2, and κ_3 are the volume suscepti-bilities of the central sample, the glass and the annular sample, respectively; a, b, and r are the inner radius of the central tube, the outer radius of the central tube, and the mean radius of the annulus, respectively. When the reference compound is in the annular space Eq. (5.4) reduces to

$$\Delta v = A\kappa_1 + C \tag{5.12}$$

where A and C are constants characteristic of the sample cell and the reference compound. A and C are best determined experimentally by measuring Δv for a series of solutions of known susceptibilities.

5.11 Specialised techniques

There are a number of more specialised techniques in high resolution n.m.r. spectroscopy which might be needed occasionally. Brief information about some of these is given in the following sections; fuller details may be found in the references quoted.

Determination of purity[6]

N.m.r. spectra of a non-spinning sample of the material are obtained over a temperature range from about 20 °C below the melting point up to a tem-perature at which the sample is completely molten. At each temperature the fraction of the sample melted is obtained from the measured intensity of the line in the spectrum arising from the liquid. The graph of temperature against reciprocal of fraction melted is linear for systems which do not form solid solutions; the slope of the graph gives the purity of the sample, and the intercept on the temperature axis gives the melting point of the pure material. A pulsed n.m.r. method for determining melting points has been described.

Optically active solvents[7]

There have been a number of reports showing that the n.m.r. spectra of enantiomers are distinguishable when an optically active compound is used as solvent. For example, in solvent fluorotrichloromethane the resonance of the CH(OH) proton in racemic phenylisopropylcarbinol

$$C_6H_5-\overset{\overset{\displaystyle OH}{|}}{\underset{\underset{\displaystyle H}{|}}{C}}-\overset{\overset{\displaystyle CH_3}{|}}{\underset{\underset{\displaystyle CH_3}{|}}{CH}}$$

is a doublet ($J = 6.8$ Hz).

When D-α-(1-naphthyl) ethylamine ($[\alpha]_D^{25} - 80.3$) is used as solvent it will associate with the two enantiomers to give collision complexes which are diastereomeric, so the carbinyl protons in the two enantiomers give two sets of doublets ($J = 6.3$ Hz) separated by 1.6 Hz at 60 MHz and 2.5 Hz at 100 MHz. Intensity measurements enable the optical purity of the solute to be determined. If the racemic amine is used as solvent only one doublet is observed from the carbinyl proton.

Liquid crystal solvents[8]

The solvents most commonly used in n.m.r. spectroscopy do not give preferred orientation of molecules when placed in a magnetic field. The chemical shifts and coupling constants obtained from the n.m.r. spectra of compounds dissolved in such a solvent are isotropic, i.e. are independent of the orientation of the molecules with respect to the magnetic field. It has been found that a number of liquid crystals can bring about partial orientation of solute molecules without excessive line broadening. The spectra obtained from partially oriented molecules can be used to study the anisotropy in chemical shifts and coupling constants, and lead to values of relative bond lengths, bond angles, and signs of spin-spin coupling constants. Two reviews of work in this field have been published[8], one deals with liquid crystals as solvents, while the second deals with liquid crystals and other methods of orientation of molecules.

Measurement of relaxation times[9]

Methods of measuring relaxation times are described in the book by Andrew. These data can be of importance in kinetic studies and in the investigation of molecular motion. See also Chapter 7.

Pulse techniques and Fourier transform spectroscopy[10]

In these techniques, instead of using a steady r.f. field, the sample is irradiated with pulses of r.f. energy. Spin-echo methods can be used for the measurement of relaxation times, and are important in kinetic studies.

When the sample is irradiated with a pulse of r.f. energy, the signal obtained is complex and depends on the absorption frequencies of all resonant nuclei in the sample. By use of the mathematical technique known as Fourier transformation, the normal absorption spectrum can be computed from the observed complex signal. Time averaging of the signal can be used, as a Fourier transform spectrum can be scanned in seconds (rather than minutes as for an ordinary n.m.r. spectrum) and the technique is particularly useful when extensive spectral accumulation has to be carried out. More details of this technique are given in Chapter 7.

Dynamic nuclear polarisation[11]

This is sometimes referred to as the electron-nuclear Overhauser effect. In solutions containing small amounts of paramagnetic free radicals, dramatic changes in intensities of n.m.r. spectral lines may be obtained by simultaneously saturating the electron spin transitions. It may be possible to estimate mean distances between interacting spins from relaxation times.

Nuclear Overhauser effect[12]

Saturating the resonance of one nucleus can affect the relaxation of another nucleus, thereby causing intensity changes in the n.m.r. spectrum. The use of this technique to determine configurations and conformations of molecules is discussed in Section 6.6.

Measurement of the strength of the radio-frequency field[13]

In certain double resonance work and in studies of relaxation behaviour it may be important to measure the strength of the radio-frequency field. A method described by Anderson makes use of modulation sidebands. The width of splitting produced in double resonance tickling experiments can be used to measure the strength of the irradiating field (Section 6.2).

References

1. VAN GEET, A. L., *Analytical Chemistry*, **40**, 2227 (1968)
2. DUERST, L., and MERBACH, A., *Rev. Sci. Inst.*, **36**, 1896 (1965)
3. Reported in Ref. 1, see also NEUMAN, R. C., and JONAS, V., *J. Amer. Chem. Soc.*, **90**, 1970 (1968)
4. MCFARLANE, W., *Chem. in Britain*, **5**, 142 (1969)
   BALDESCHWIELER, J. D., and RANDALL, E. W., *Chem. Rev.*, **63**, 81 (1963)
   MCFARLANE, W., *Ann. Rev. of N.m.r. Spectro.*, **1**, 135 (1968)
   HOFFMAN, R. A., and FORSEN, S., *Prog. in N.m.r. Spectro.*, **1**, 15 (1966)
5. EVANS, D., *J. Chem. Soc.*, 2003 (1959)
   DEUTSCH, J. L., LAWSON, A. C., and POLING, S. M., *Analytical Chemistry*, **40**, 839 (1968)
   DEUTSCH, J. L., and POLING, S. M., *J. Chem. Ed.*, **46**, 167 (1969)

6. HERINGTON, E. F. G., and LAWRENSON, I. J., *Nature*, **219**, 928 (1968)
   BURNETT, L. J., and MULLER, B. H., *Nature*, **219**, 59 (1968)
7. PIRKLE, W. H., and BEARE, S. D., *J. Amer. Chem. Soc.*, **89**, 5485 (1967) and references therein
8. LUCKHURST, G. R., *Q. Rev. Chem. Soc.*, **22**, 179 (1968)
   BUCKINGHAM, A. D., and MCLAUCHLAN, K. A., *Prog. in N.m.r. Spectro.*, **2**, 63 (1967)
9. ANDREW, E. R., *Nuclear Magnetic Resonance*, Cambridge (1955)
10. HAHN, E. L., *Phys. Rev.*, **80**, 580 (1950)
    CARR, H. Y., and PURCELL, E. M., *Phys. Rev.*, **94**, 630 (1954)
    ERNST, R. R., and ANDERSON, W. A., *Rev. Sci. Instr.*, **37**, 93 (1966)
    FARRAR, T. C., *Analytical Chemistry*, **42**, 109A (1970)
11. ANDERSON, W. A., *N.m.r. and E.p.r. Spectroscopy*, Pergamon, 268 (1960)
    RICHARDS, R. E., and WHITE, J. W., *Disc. Faraday Soc.*, **34**, 96 (1962)
12. KENNEWELL, P. D., *J. Chem. Ed.*, **47**, 278 (1970)
13. ANDERSON, W. A., *N.m.r. and E.p.r. Spectroscopy*, Pergamon, 164 (1960)

6   Double resonance techniques

6.1 Introduction

One of the simplest and most generally valuable applications of double resonance, namely complete decoupling, has been dealt with in Section 3.2; but there are a number of other and more sophisticated uses of this technique. These include spin tickling for the measurement of 'hidden' splittings, and the determination of their relative signs; the determination of the chemical shifts of nuclei with low inherent sensitivity to n.m.r. detection; and Overhauser experiments which may be used for the direct study of molecular geometry[1, 4]. The basic principles behind these techniques and their application to a number of simple illustrative systems will now be described.

6.2 Spin tickling

If the amplitude of the r.f. field used for irradiation in a normal decoupling experiment (p. 71) is too low, then quite complicated spectra may result, and in general these are rather difficult to interpret. However, if the amplitude is very low indeed then the observed effects can be described simply, and the term 'tickling' is often applied to such experiments. The results are most straightforward when the spectra are recorded by sweeping frequency, for it is then easy to maintain a fixed relationship between the setting of the irradiating frequency and the positions of lines in the spectrum of the irradiated nucleus. For such experiments it is then found[5] that:

(a) No effects are observed unless the frequency of the irradiating field is close to that of a *line* in the spectrum of the irradiated nucleus
(b) When the frequencies of the irradiating field and of a line in the irradiated spectrum correspond exactly, certain lines in the observed spectrum will be split into symmetrical doublets. The lines which are split will be those which have an energy-level in common with the irradiated line
(c) The width of the splitting produced in this way depends both upon the strength of the r.f. field and the intensity of the line which is irradiated. For first-order spectra the second factor may be ignored and the

width of the splitting in Hz is then $\gamma B_2/2\pi$, where γ is the magnetogyric ratio of the irradiated nucleus and B_2 is the amplitude of the irradiating field. This relationship can actually be used to measure the strength of the irradiating field, and the condition that this field be *weak* is really that $\gamma B_2/2\pi$ should be much smaller than the separation between any pair of lines in the spectrum of the irradiated nucleus. (This can be taken as a definition of tickling)

(d) If the irradiating r.f. field is set a short way off exact resonance with a line, then the same lines will still be split into doublets, but these will now be unsymmetrical. Thus, by following the changes in the heights of the components of one of the doublets it is possible to adjust the frequency of the irradiating field until exact coincidence with a line is achieved.

The application of these rules is illustrated in Figure 6.1, which shows double resonance spectra of the phosphite anion HPO_3^{-} $^{-}$ in which there is a direct P–H bond with $^1J(^{31}P-H) = 580$ Hz. The normal proton

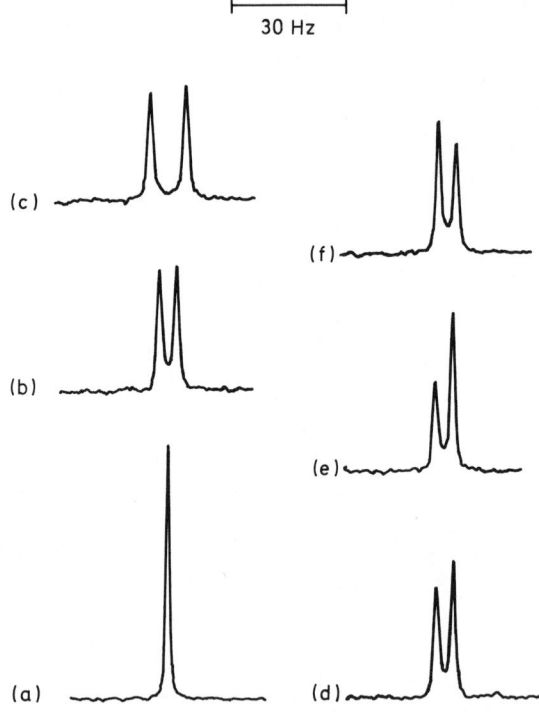

Figure 6.1. Low-field component of the doublet in the proton spectrum of the HPO_3^{-} $^{-}$ anion. (a) Unperturbed; (b) with simultaneous irradiation at a phosphorus frequency of 24 286 343 Hz with amplitude given by $\gamma B_2/2\pi = 5$ Hz; (c) same frequency, amplitude 10 Hz; (d) frequency 1 Hz lower, amplitude 5 Hz; (e) frequency 24 286 339 Hz, amplitude 5 Hz; (f) frequency 24 286 344, amplitude 5 Hz

spectrum is a doublet of separation 580 Hz, one component of which is shown in Figure 6.1 (a), and the ^{31}P spectrum will be identical to this, with lines at 24 286 343 and 24 286 923 Hz at this particular magnetic field strength.

If a weak field (actual amplitude given by $\dfrac{\gamma B_2}{2\pi} = 5$ Hz) is applied to the sample at a frequency of 24 286 343 Hz (this corresponding to the *high-field* ^{31}P line; note the inverse relationship between field and frequency in this context) each of the lines in the proton spectrum is split into a symmetrical doublet as shown in Figure 6.1 (b). If the intensity of the second irradiating field is altered then the width of the splitting will also change (Figure 6.1 (c)), while if the frequency of the irradiating field is changed to 24 286 342 Hz, i.e. offset by 1 Hz from exact resonance the doublets in the proton spectrum become asymmetrical as shown in (d). Increasing the frequency offset increases the asymmetry (Figure 6.1 (e)), while a similar frequency offset in the opposite direction reverses the sense of the asymmetry (Figure 6.1 (f)). Similar results can be obtained by irradiating in the neighbourhood of the other ^{31}P line at 24 286 923 Hz, and it would be possible to determine the positions of these phosphorus lines to ± 0.1 Hz by using lower amplitudes of the irradiating field so as to give smaller splittings. Finally, irradiation at a frequency of 24 286 633 Hz (corresponding to the centre of the ^{31}P spectrum, although there is no line at this point) with a weak field has scarcely any effect, although a sufficiently intense field would have given complete decoupling.

Usually the spin system under investigation is not as simple as the AX one considered above, and it is then common to find that only a portion of an observed line is split into a doublet in a particular tickling experiment[6]. The remainder is unaffected and so a symmetrical *triplet* results. The intensity pattern in the triplet will depend upon the intensity of the line that is irradiated, and for $A - \{X\}^*$ experiments on an $A_n X$ spin system, the proportion of the observed line which is split into a doublet is equal to twice the fractional intensity of the irradiated line. For example in an $A_3 X$ system irradiation of either of the central components of the X $1:3:3:1$ quartet will affect three-quarters of each component of the A doublet to give a $3:2:3$ pattern for each original line, and irradiation of either of the outer X lines will give $1:6:1$ triplets. This is illustrated in Figure 6.2 which shows the high-field ^{13}C satellite of the proton spectrum of methyl iodide under various conditions of ^{13}C irradiation. Other simple systems can be treated similarly, and there is normally no particular difficulty in determining the positions of individual lines in the spectrum of the irradiated nucleus to ± 0.1 Hz. However, the reader is warned that the off-resonance behaviour of these systems is not as simple as might be expected.

It has been assumed in the foregoing that the overall intensity of lines in the observed spectrum remains unaffected by a double resonance experiment. This may not necessarily be so, although when the observed nucleus has a magnetogyric ratio considerably larger than the irradiated one the effects are quite small. In homonuclear and certain other experiments, however, quite large effects are found and these are due to selective spin saturation of

---

* This symbol indicates that the resonance of A is observed while X is irradiated.

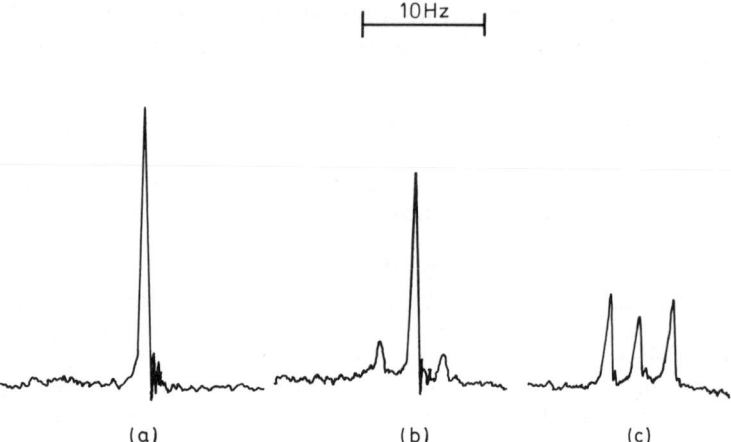

Figure 6.2. High-field ^{13}C satellite in the proton spectrum of methyl iodide. (a) Un-perturbed line; (b) with irradiation at the frequency of the line of the ^{13}C quartet at lowest field: $\frac{1}{4}$ of the intensity of the recorded line is affected; (c) with irradiation at the frequency of one of the central lines of the ^{13}C quartet: $\frac{3}{4}$ of the recorded line is affected

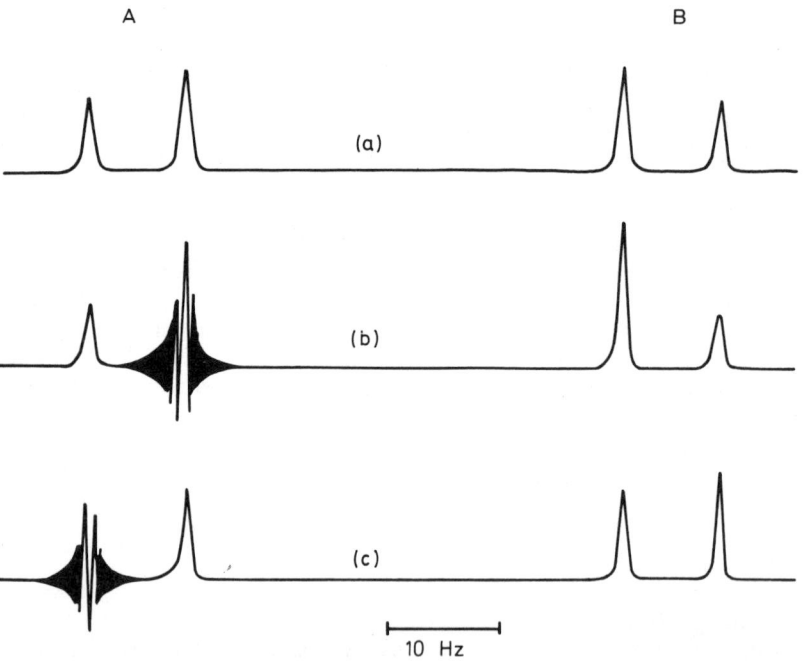

Figure 6.3. Double resonance experiments on an AB spin system. (a) Normal spectrum; (b) irradiation of the high-field A line increases the intensity of the low-field B line and diminishes the high-field B line; (c) irradiation of the low-field A line produces the opposite effects. The amplitude of the irradiating field in these experiments is given by $\gamma B_2/2\pi = ca\ 0.1$ Hz, i.e. it is just too weak to produce resolvable splitting

the line which is irradiated. That is, the spin populations of the energy-levels connected by the irradiated transition become equalised, and this in turn affects the intensities of lines in the spectrum of the observed nucleus. These effects are illustrated in Figure 6.3 for a simple AB spin system using amplitudes of the irradiating field that are just too small to give resolved line splittings. The position of the beat pattern indicates the setting of the tickling field, and it is clear that when the high-field component of the A doublet is irradiated the low-field line in the B spectrum increases in intensity and the high-field one decreases. Irradiation of the low-field A line has the converse effect, and this type of relationship applies in more complicated systems. The behaviour can be predicted by considering the details of the energy-level diagram[7].

Figure 6.3 also shows that the intensity changes are accompanied by differential broadening effects. These depend upon the magnetic field in homogeneity, and in general those lines which increase in intensity also become broader, while those which lose intensity become sharper than in the normal spectrum. The greatest precision is attained in tickling experiments by observing those lines which are sharpened.

6.3 The determination of chemical shifts

It should be clear from the foregoing that the details of the fine structure of the spectrum of a heteronucleus may be determined by means of a series of tickling experiments, and the data so obtained can be used to give extremely accurate chemical shifts. Furthermore, these shifts can be related to an internal proton reference, and so detailed solvent studies may be easily performed. For purposes of comparison it is desirable to quote resonant frequencies obtained at some definite magnetic field strength, and a convenient standard is *that field in which T.M.S. gives a proton resonance of exactly 100 MHz*[3]. Apart from a factor of 10^8 this is equivalent to quoting the ratio of the resonant frequencies of the nucleus and of T.M.S. in the same magnetic field[5]. Normally, the field actually used for the measurement will not correspond precisely to this frequency, and in this case the observed frequency may be corrected by using Eq. (6.1).

$$\Xi = X_{obs.} + X_{obs.} (s - f + 100\delta)/10^8 \qquad (6.1)$$

Where Ξ is the required frequency, $X_{obs.}$ is observed frequency in Hz, s is the frequency of the audio oscillator used to provide the locking signal (see p. 72), f is the amount by which the actual centreband frequency of the spectrometer exceeds 100 MHz, and δ is the chemical shift to low field of T.M.S. of the signal used to actuate the field-frequency locking circuit. This formula applies when the low-field modulation sideband is used for field-frequency locking; if the high-field sideband is used the sign of s must be reversed. Similar equations can be written to correct values obtained on spectrometers which operate at proton frequencies of 60 or 90, and division by 0.6 or 0.9 will then lead to the corresponding Ξ values.

The approach described above can give chemical shifts to better than ± 0.01 p.p.m., but for many purposes this degree of precision is unnecessary, and a simpler procedure will suffice. In a reasonably stable environment the frequency of the spectrometer oscillator is unlikely to vary by more than 30 to 40 Hz (corresponding to 0.3 to 0.4 p.p.m. at 100 MHz) in the course of a few hours, so comparative measurements may be undertaken, provided that the resonant frequency of a suitable reference compound is determined occasionally. Furthermore, it is often quite sufficient to measure the frequency which gives optimum decoupling of the heteronucleus from the protons, and to operate in the field-sweep mode. In this case, the chemical shift of the proton resonance that is actually observed must be taken into account in determining the required chemical shift. For example, $^1H - \{^{31}P\}$ field-sweep experiments on a spectrometer operating at about 60 MHz for protons on triphenyl phosphine and trimethyl phosphite, gave 24 286 335 and 24 290 034 Hz respectively for the ^{31}P resonant frequencies in the two compounds. The δ-value of triphenyl phosphine is 7.2 so by adding $7.2 \times 24.28 = 175$ Hz to the first frequency, we get the ^{31}P resonant frequency of triphenyl phosphine in the field appropriate to a T.M.S. proton resonance at the *actual operating frequency of the particular spectrometer used*. The δ-value for trimethyl phosphite is 3.68 so the correction is 89 Hz, and the corrected ^{31}P resonance frequencies in the two compounds are 24 286 510 and 24 290 125 Hz. These figures are directly comparable since they were obtained on the same spectrometer with only a short time interval between the two measurements. Drift of the spectrometer oscillator should thus be negligible. It is important to remember that in this connection, a higher resonant frequency indicates a shift to lower field, so the experiments show that triphenyl phosphine gives a ^{31}P resonance 3615/24.29 = 148.4 p.p.m. to high field of trimethyl phosphite.

6.4 Determination of the relative signs of coupling constants

A coupling constant is essentially a measure of the extent to which the spin orientation of one nucleus can affect that of another. This interaction may be negative or positive, that is the energy of a nuclear spin may be lowered *either* by a second nucleus with parallel spin *or* by a second nucleus with anti-parallel spin. For nuclei with a positive magnetogyric ratio the latter situation is said to give a positive coupling constant. A reversal of *all* the signs of the coupling constants in a spin system has no effect upon the appearance of the spectrum or upon the results of double resonance experiments, and consequently it is extremely difficult to determine the *absolute* signs of coupling constants.* However, if the signs of some (but not all) of the coupling constants are reversed this may alter the appearance of second-order spectra (p. 12) and will probably affect the results of double resonance experiments. It is thus easy to determine the *relative* signs of coupling constants, and provided that the absolute signs of a few *key* coupling constants are known, the signs of many others may be related to these.

The results of sign-determining double resonance experiments may be

---

* Methods for doing this generally depend upon studying molecules which are *partially oriented*, either by application of a strong electric field or by dissolution in a liquid crystal solvent.

interpreted in several ways, but we shall confine ourselves to the sub-spectral approach. This may be illustrated conveniently by the olefinic protons of vinyl acetate:

$$\underset{H_X}{\overset{AcO}{\diagdown}}C = C\underset{H_M}{\overset{H_X}{\diagup}}$$

Figure 6.4 (a) shows the spectrum given by H_A and H_M; the narrow doublet splitting is due to the coupling (1.4 Hz) between these two protons. This spectrum may be divided into two *sub-spectra*, each associated with a

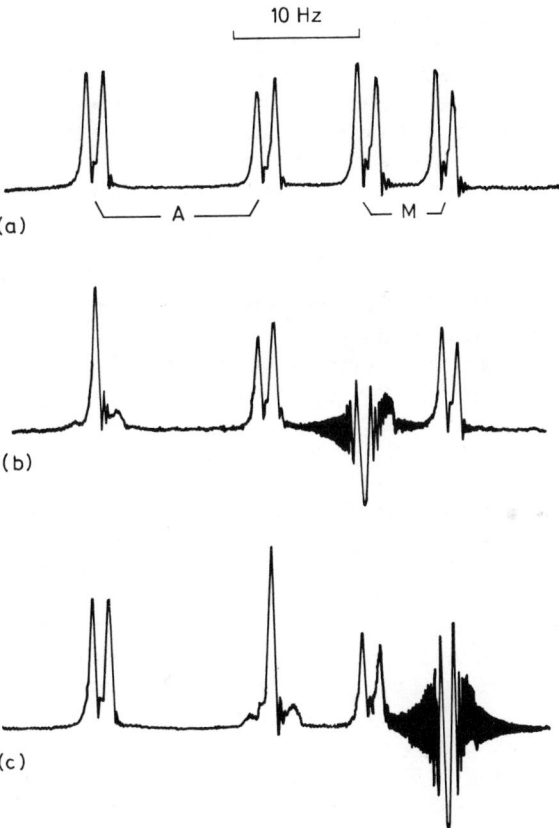

Figure 6.4. Selective decoupling experiments on the AMX spin system given by the olefinic protons of vinyl acetate. (a) Normal spectrum of H_A and H_M; (b) with irradiation of the low-field H_M doublet, the low-field H_A doublet is decoupled; (c) with irradiation of the high-field H_M doublet, the high-field H_A doublet is decoupled

different spin orientation of H_X, and the question of which doublets arise from a particular spin state depends upon the signs of J_{AX} and J_{MX}. Specifically, if these two signs are the same then the high-field H_A and the high-field

H_M doublets will belong to the same sub-spectrum, whereas if the two signs are opposite then these doublets will belong to different sub-spectra. If the low-field H_M doublet is irradiated, the H_A doublet *belonging to the same sub-spectra* (i.e. the one that is associated with the same spin orientation of H_X) will be decoupled, but the H_A doublet belonging to the other sub-spectrum will be unaffected. This is illustrated in Figure 6.4 (b) which shows that the low-field H_A and H_M doublets belong to the same sub-spectrum, so that J_{AX} and J_{MX} are of like sign. In Figure 6.4 (c) this conclusion is confirmed by irradiating the high-field H_M doublet, when it is found that the high-field H_A doublet is decoupled. This method for determining the relative signs of coupling constants is known as *selective decoupling*, and its first application was by Evans[8] who used it to show that $^2J(^{205}Tl...CH_2)$ and $^3J(^{205}Tl...CH_3)$ are of opposite sign in the diethyl thallium cation.

In general the foregoing technique can be used whenever the coupling that is to be removed is small compared with the other spacings in the spectrum of the irradiated nucleus. That is, if the signs of J_{AX} and J_{MX} in an AMX spin system are to be compared, then J_{AM} must be considerably smaller than either of the other two coupling constants. A practical point here is that the amplitude of the decoupling field should be set rather lower than would be usual in an experiment intended to simplify a spectrum by de-coupling. Also the field is more intense than that used for tickling. It is more important to avoid perturbing the wrong lines than to achieve *complete* collapse of a particular coupling.

It should now be clear that attempts to compare the sign of J_{AM} with that of J_{AX} or J_{MX} in vinyl acetate, by selective decoupling, would fail because in the first case it would be necessary to collapse J_{MX} (8 Hz), and in the second case J_{AX} (15 Hz), and both of these couplings are much larger than J_{AM} itself. Instead, it is necessary to have recourse to tickling experiments in which *individual lines* are perturbed. In Figure 6.5 (c) the effect of tickling the H_X line at lowest field is shown. The amplitude of the irradiating field is sufficient to give splittings of 1 Hz of *connected* lines (see p. 85), and so one component of each of the H_A and H_M doublets suffers this splitting. *The affected lines belong to the same sub-spectrum as the irradiated one*, but the choice of sub-spectra will depend upon which signs are to be compared. Thus, if the signs of J_{AM} and J_{MX} are to be compared, then the sub-spectra corresponding to opposite spin states of H_M must be considered. The two H_X lines at lowest field are separated by J_{MX} and, therefore, are associated with opposite spin orientations of H_M, i.e. they belong to different sub-spectra; similarly the two components of either H_A doublet belong to different sub-spectra. The experiments illustrated in Figures 6.5 (b) and (c) show that a *high-field* H_X line is associated with a *low-field* H_A one, and *vice versa*, so J_{AM} and J_{MX} are of opposite sign.

In order to get the relative signs of J_{AM} and J_{AX} we must consider the two sub-spectra associated with opposite spins of H_A. The two lines at lowest field in the spectrum of H_X are associated with the *same* spin orientation of H_A, that is they belong to the same sub-spectrum, so irradiation of either of these lines should split the same lines in the H_M spectrum. Figures 6.5 (b) and (c) show that this is indeed so, although there is a small difference in the overall appearance in the two cases owing to the reversal of the progressive-regressive arrangement (see ref. 3). Irradiation of either of the two H_X lines at higher field will affect the other pair of H_M lines (this experiment is not actually shown in the figure), and since these are at *lower field* it follows that J_{AX} and J_{AM} are of opposite sign. It will be noticed that the results of these three sign determinations are fully self-consistent, and in fact only two experiments need actually be performed although the third provides a useful check. The relative signs are thus: $J_{AM}\mp$; $J_{AX}\pm$; $J_{MX}\pm$; and in fact it is known from other studies that the upper set of signs is the correct one. In general then, two distinct double resonance experiments are necessary for

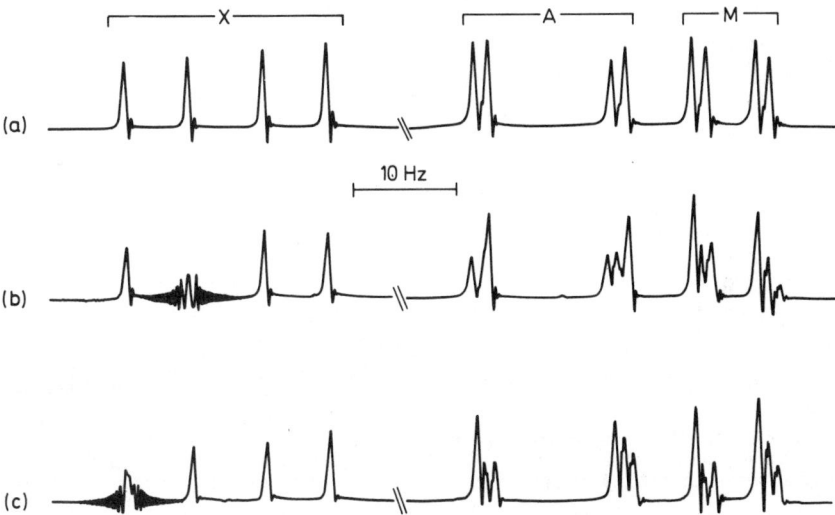

Figure 6.5. Spin tickling experiments on the AMX system given by the olefinic protons of vinyl acetate. (a) Normal spectrum; (b) with irradiation of $H_X(2)$: transitions $H_A(1)$, $H_A(3)$, $H_M(2)$, and $H_M(4)$ are split into doublets while the others are unaffected; (c) with irradiation of $H_X(1)$: transitions $H_A(2)$, $H_A(4)$, $H_M(2)$, and $H_M(4)$ are split

the determination of all the relative signs in an AMX spin system, and each nucleus must be either irradiated or observed in at least one of the two experiments. The extension to more elaborate spin systems (e.g. AMPX etc.) is straightforward, and usually involves a breakdown into a greater number of sub-spectra; however, if second-order features are present it may be necessary to consider the details of the appropriate energy-level diagram. The question of whether to use selective decoupling or tickling experiments can be decided only by consideration of the details of each problem; the former approach puts smaller demands upon instrumental stability, and the results may be interpreted rather more easily.

The techniques described above can be applied with equal facility to heteronuclear systems. It is normally possible to confine observation to one nuclear species (usually the proton or ^{19}F), and still get the relative signs of *all* the coupling constants by suitable heteronuclear double resonance experiments. These experiments will also give the *magnitudes* of the coupling constants that do not involve the observed nuclei directly. A typical example of this method is illustrated in Figure 6.6 which shows frequency-sweep 60 MHz proton spectra of trimethyl phosphate, $(CH_3O)_3PO$. The main resonance is a doublet owing to splitting by ^{31}P, and is flanked by ^{13}C satellites (see p. 31) which arise from molecules of the type $(CH_3O)_2(^{13}CH_3O)PO$. The ^{13}C

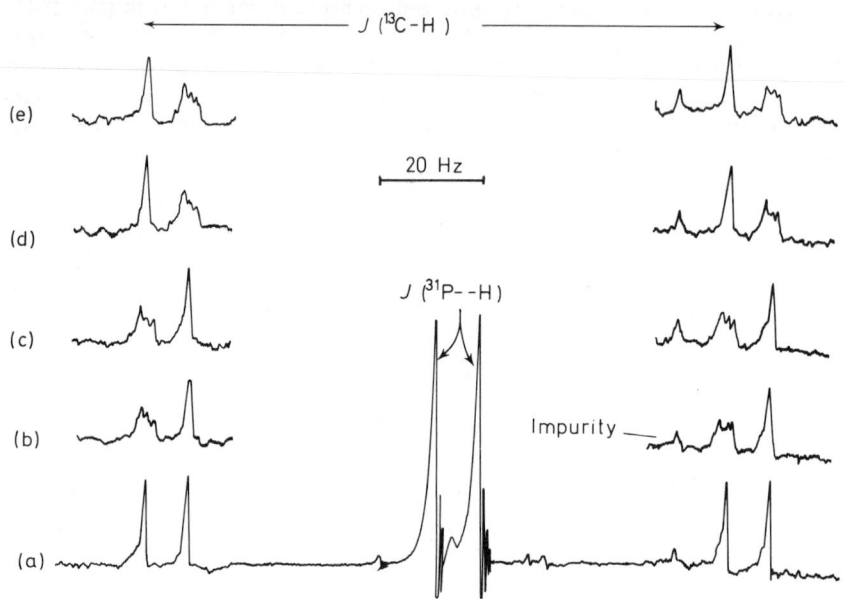

Figure 6.6. $^1H-\{^{13}C\}$ double resonance experiments on trimethyl phosphate, $(CH_3O)_3PO$. (a) Normal proton spectrum showing ^{13}C satellites at increased spectrometer gain; (b) with irradiation at 15 086 631.4 Hz; (c) 15 086 778.4 Hz; (d) 15 086 637.2 Hz; (e) 15 086 784.2 Hz

spectrum of this kind of molecule will be two $1:3:3:1$ quartets of internal spacing $^1J(^{13}C-H) = 147$ Hz and separated by $^2J(^{13}C...^{31}P)$. The positions of the components of one of these quartets can be determined by observing the high-field part of either ^{13}C satellite, and performing $^1H-\{^{13}C\}$ tickling experiments, some of which are illustrated in Figures 6.6 (d) and (e). The positions of the components of the other ^{13}C quartet can be determined similarly by observing the low-field half of either ^{13}C satellite (Figures 6.6 (b) and (c)) and the results of these experiments are summarised below.

	High-field ^1H lines	Low-field ^1H lines
Associated ^{13}C frequencies	$15\,086\begin{cases}784.2\\637.2\end{cases}$ Hz	$15\,086\begin{cases}778.4\\631.4\end{cases}$ Hz
Mean ^{13}C frequency	15 086 710.7 Hz	15 086 704.9 Hz

The differences between corresponding ^{13}C frequencies is 5.8 Hz, so this is the magnitude of $^2J(^{13}C...$ $^{31}P)$, and since the *lower* ^{13}C *frequencies* (i.e. corresponding to resonances at *higher field*) are associated with the proton lines at *lower field*, it follows that $^2J(^{13}C...^{31}P)$ and $^3J(^{31}P...H)$ are of opposite sign.

The ^{31}P spectrum is a binomial decet with an internal spacing of 10.0 Hz and is flanked by weak overlapping ^{13}C satellites (these are also decets). The position of the centreband was determined by observing the main proton doublet and performing $^1H - \{^{31}P\}$ tickling experiments, and was found to be centred on 24 286 746 Hz. Similarly the ^{31}P decet associated with the low-field ^{13}C satellite in the proton spectrum was at 24 286 743 Hz, and the remaining decet was at 24 286 749 Hz. These values confirm the magnitude of $^2J(^{13}C \ldots ^{31}P)$ and show that it is of opposite sign to $^1J(^{13}C - H)$. This latter coupling constant is known to be absolutely positive, so the coupling constants in trimethyl phosphate are

$$^1J(^{13}C - H) = +147.0 \text{ Hz}, \quad ^2J(^{13}C \ldots ^{31}P) = -5.8 \text{ Hz},$$

$$^3J(^{31}P \ldots H) = +10.0 \text{ Hz}$$

6.5 INDOR spectra

One of the disadvantages of the double resonance methods outlined above for obtaining the details of the spectrum of a nucleus that is not observed directly is that they are rather time consuming. This difficulty is avoided to a considerable extent by the INDOR (Inter Nuclear Double Resonance) technique in which the height of a suitable line in the proton spectrum is

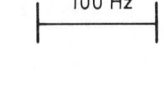

100 Hz

Figure 6.7. ^{195}Pt INDOR spectrum of trans $[(CH_3)_2S]_2PtCl_2$ obtained by monitoring a ^{195}Pt (abundance 34%) satellite in the proton spectrum. Seven of the expected thirteen lines are clearly recorded

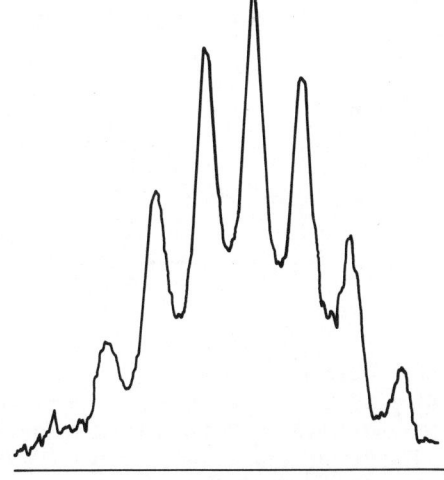

monitored continuously, and variations are recorded as the second r.f. field is swept through the region of interest[9]. The precise form of these variations will depend upon the amplitude of the irradiating field, the spacings in the irradiated spectrum, certain relaxation times, and other factors. However, if it is possible to have the second r.f. strong enough to give line splittings of several hertz without seriously affecting more than one line of the irradiated spectrum at a time, then the output of the recorder may resemble the 'normal' spectrum of the irradiated nucleus quite closely. Difficulties tend to arise when the amplitude of the irradiating field is quite low, and the effects of spin population transfer may then predominate. In these circumstances the recorded spectrum may have lines on either side of the baseline, and it is difficult to interpret the relative intensities satisfactorily. Figure 6.7 shows the ^{195}Pt INDOR spectrum of trans$[(CH_3)_2S]_2PtCl_2$ under conditions which minimise these difficulties, and it will be noticed that resemblance to a binomial multiplet is quite good. Figure 6.8 shows ^{19}F INDOR spectra

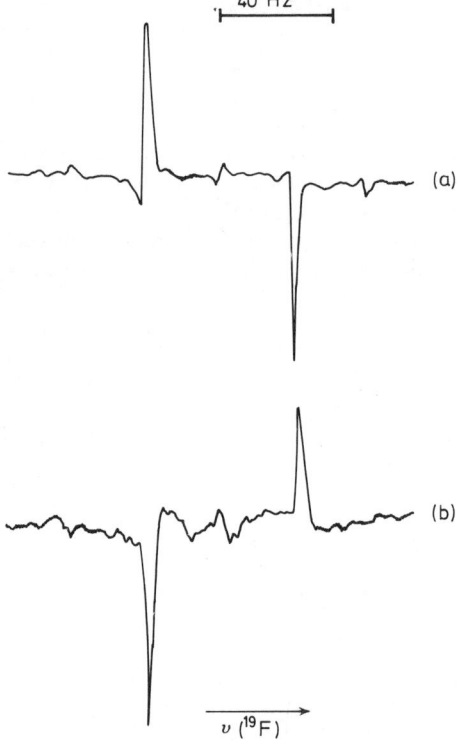

Figure 6.8. ^{19}F INDOR spectra of difluoracetic acid in which $J_{HF} = 52$ Hz. (a) Low-field line in proton spectrum monitored; (b) high-field line in proton spectrum monitored

of CHF_2COOH, obtained by monitoring different lines in the proton spectrum. Here the main effects are due to spin population transfer, which can either diminish or augment the intensity of the monitored line, so that the recorded trace shows both negative and positive peaks.

The production of high-quality INDOR spectra places high demands upon the stability of the spectrometer, and an internal field-frequency locking system is almost essential. The second frequency may be driven through the required range either by mechanical or electrical means, and if the latter method is used a direct link to the recorder can then give pre-calibrated INDOR spectra. INDOR spectra are of greatest value when the nucleus of interest has a low magnetogyric ratio, for then the gain in sensitivity will be greatest. Furthermore, for these nuclei the effects of spin population transfer will be relatively slight, so that the resemblance to the normal single resonance spectra will be close. If the observed and irradiated nuclei are both coupled to other nuclei then several INDOR spectra will be obtained, according to which line is monitored. The complete spectrum is a superposition of these, although some care must be taken in assessing the relative intensities.

6.6 The nuclear Overhauser effect

It has been pointed out earlier that many double resonance experiments are accompanied by changes in the overall intensities of the observed lines. These arise from changes in the population of the energy-levels which result from saturation of an irradiated transition by the second r.f. field. This phenomenon has been referred to as the *Generalised Nuclear Overhauser Effect*, but this designation has the disadvantage that it can lead to confusion with the true *Nuclear Overhauser Effect* (NOE) which arises when saturation of the resonance of one nucleus can effect the *relaxation* of another, and so produce intensity changes in this way[2]. This latter phenomenon occurs when the nuclei in question are physically close, although there need not necessarily be any spin coupling between them. It is then found that the relaxation of the observed nuclei may depend mainly upon spin exchange with the irradiated one, so that saturation of the latter can give an intensity enhancement of up to 50% when the two nuclei are of the same species. The requirement that the two nuclei be close enough for this type of relaxation to be important really demands that their separation should not exceed the sum of the two van der Waals' radii—*ca* 200 m^{-12} for protons—and the effect therefore can be used in the study of steric overcrowding. Intensity enhancements that approach the theoretical maximum can be expected only when other relaxation processes are negligible, and this is rather difficult to achieve with protons as these normally lie on the periphery of a molecule. In particular it is necessary to use a fairly dilute solution in a solvent that has no nuclei of high magnetic moment. This precludes the use of solvents containing protons or ^{19}F nuclei, and also rules out other halogenated materials. Carbon disulphide has been used with success, and the magnetic moment of the deuteron is low enough for its presence not to be serious so that various deuteriated solvents are also suitable. In addition, the removal of dissolved oxygen (this is paramagnetic) may be advantageous. An example of this technique is shown in Figure 6.9. The lower trace shows the normal spectrum of the aromatic protons of 2-methoxy 5-hydroxy benzaldehyde in which $J(H_{(3)} - H_{(4)}) = 7.2$ Hz and $J(H_{(3)} - H_{(6)}) = 0$ Hz, so that $H_{(3)}$ gives the doublet at high field. Simultaneous irradiation of the methoxy line enhances the intensity of the $H_{(3)}$ by *ca* 15%, but leaves the $H_{(4)}$ and $H_{(6)}$ resonances essentially unaffected, as shown in the upper trace. Steric effects

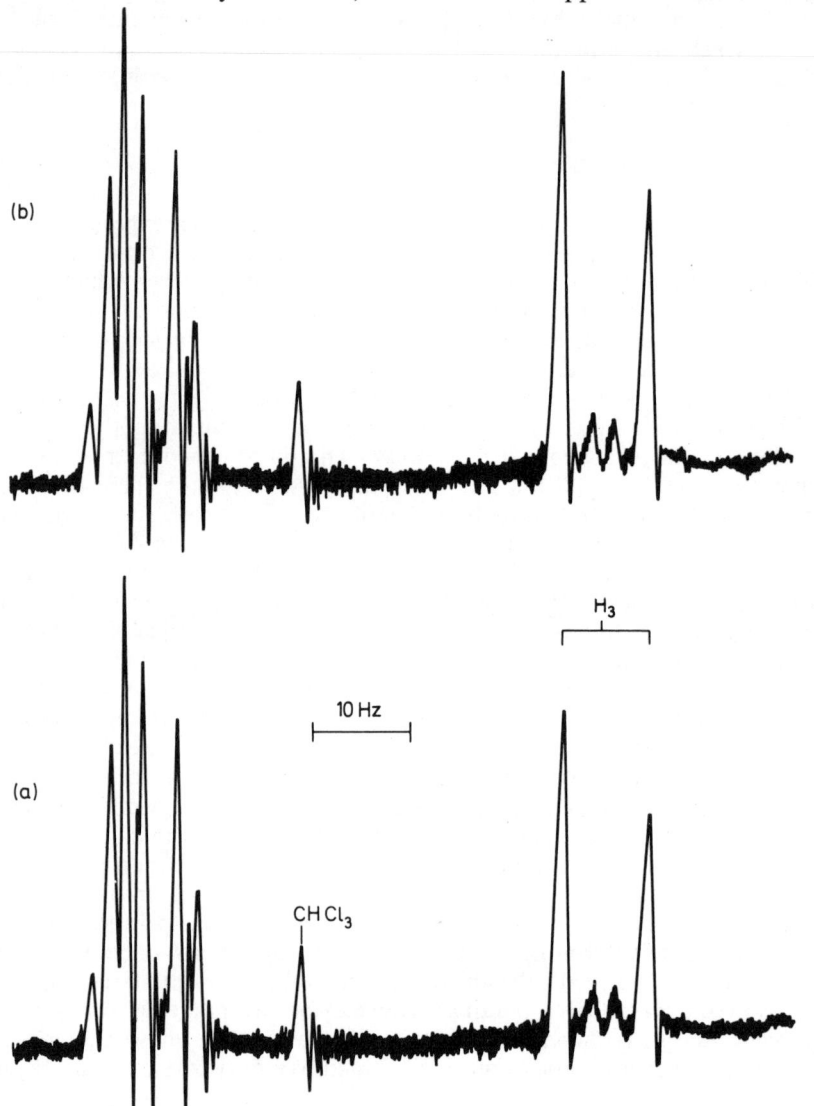

Figure 6.9. Proton spectra (at 100 MHz) of the aromatic protons of 2-methoxy 5-hydroxy benzaldehyde. (a) Normal spectrum; (b) with simultaneous irradiation at the frequency of the methoxy resonance. (Courtesy of Japanese Electron Optico Laboratory)

involving the aldehyde and methoxy groups constrain the methyl group to a position close to $H_{(3)}$ so that there is a substantial nuclear Overhauser effect.

The foregoing experiment was carried out under steady state conditions; that is one resonance was saturated *continuously* while the others were observed, and this is the normal way of performing NOE experiments. However, the observed intensity change can be doubled if instead of *saturating* one resonance (i.e. equalising the spin populations) the spin populations can be inverted. This can be done by making an adiabatic fast passage through the resonance to be irradiated immediately (a few seconds) prior to recording the region of interest. In many cases this can be accomplished easily by using the field-sweep mode of operation together with a setting of the irradiating frequency that is a few hertz off exact resonance, and recording the spectrum fairly rapidly. However, if the region to be irradiated contains resonances from several different protons then the results may not be unequivocal.

References

1. BALDESCHWIELER, J. D., and RANDALL, E. W., *Chem. Rev.*, **63**, 81 (1963)
2. HOFFMAN, R. A., and FORSÉN, S., *Prog. in N.m.r. Spectro.*, **1**, 15 (1966)
3. MCFARLANE, W., *Ann. Rev. N.m.r. Spectroscopy*, **1**, 135 (1968)
4. MCFARLANE, W., *Phys. Methods for Det. Org. Structures*, **4**, 139 (1971)
5. FREEMAN, R., and ANDERSON, W. A., *J. Chem. Phys.*, **37**, 2053 (1962)
6. MCLAUCHLAN, K. A., WHIFFEN, D. H., and REEVES, L. W., *Mol. Phys.*, **10**, 131 (1966)
7. KAISER, R., *J. Chem. Phys.*, **39**, 2435 (1963)
8. MAHER, J. P., and EVANS, D. F., *Proc. Chem. Soc.*, 208 (1961)
9. BAKER, E. B., *J. Chem. Phys.*, **37**, 911 (1962)
10. KENNEWELL, P. D., *J. Chem. Ed.*, **47**, 278 (1970)

7 N.m.r. spectra of carbon-13 and other nuclei

7.1 Introduction

In previous sections we have considered the methods used in hydrogen n.m.r. spectroscopy. Although many of these techniques are applicable to the study of other nuclei, differences in inherent sensitivity to n.m.r. detection, natural abundance, relaxation behaviour and overall range of n.m.r. parameters may necessitate certain alterations in technique.

For our present purposes n.m.r. spectrometers may be classified as either

    (a) instruments that can be operated at a fixed (or only slightly variable) field strength
    (b) instruments that can be operated over a wide range of field strengths.

With the former type of spectrometer, an r.f. unit operating at an appropriate frequency will be required for *each* isotope that is to be observed, whereas with the latter type of instrument one r.f. unit may be used to study several different isotopes. Normally the manufacturer will provide and line up the r.f. unit necessary to observe a particular nuclear species with a fixed field spectrometer. This system is very convenient, but the cost of the necessary units can restrict the number of other isotopes that can be studied.

An instrument with a magnet that is suitable for use over a range of field strengths provides a less expensive way of investigating a number of different isotopes; however, the operator is then faced with the difficulty of finding the appropriate field for each isotope. In this case it is useful to keep a record of the magnet current, cycling procedure (see p. 50), shim coil settings, phase adjustment, and other settings for each isotope studied. A difficulty commonly encountered in this type of work is that at higher field strengths the magnet current and the field strength are not linearly related. Appendix 5 (a) gives the resonance fields of a number of isotopes for a range of fixed frequencies; data of this sort can be used to build up a field strength/magnet current calibration graph for a particular instrument. Such a calibration graph can be of considerable help when first finding the field for a particular isotope. We now consider some of the more important points in relation to the more frequently studied nuclei other than protons,* and in particular carbon-13.

7.2 Fluorine[1]

The isotope ^{19}F occurs in 100% natural abundance, and has almost the same sensitivity to n.m.r. detection as the proton, so the experimental methods for observing these nuclei are very similar. In some laboratories it is found convenient to run both ^{1}H and ^{19}F spectra at the same frequency (say 56.4 MHz) using a reduced field for proton work. The loss in signal-to-noise ratio for the proton spectrum is insignificant, although the arithmetic of calculating δ values may be more tedious.

The main ways in which fluorine n.m.r. spectroscopy differs from proton work are:

(a) T_1, the spin-lattice relaxation time is often much longer for fluorine, so that it is harder to avoid saturation

(b) the total chemical shift range is much greater than for protons, consequently it may be less easy to record frequency-sweep ^{19}F spectra (phase adjustment may be necessary when the frequency is altered by large amounts); alternatively it may be necessary to adjust the homogeneity in the course of an extended field-sweep

(c) solvent effects upon ^{19}F chemical shifts (5 p.p.m. or more) are much larger than for protons; it is a good rule to use a reference compound that is chemically similar to the sample (e.g. C_6F_6 for fluorinated aromatics) whenever possible.

Common reference compounds used in fluorine n.m.r. spectroscopy are: $CFCl_3$ used as an internal reference, defines the ϕ scale; CF_3COOH used as an external reference, susceptibility correction is usually ignored; C_6F_6 used as an internal reference. Approximate relations† between these three reference scales are given in Equations (7.1) to (7.3)

$$\delta(\text{rel. } C_6F_6) = \delta(\text{rel. } CF_3COOH) + 84.4 \text{ p.p.m.} \tag{7.1}$$

$$\delta(\text{rel. } C_6F_6) = \delta(\text{rel. } CFCl_3) + 162.9 \text{ p.p.m.} \tag{7.2}$$

$$\delta(\text{rel. } CFCl_3) = \delta(\text{rel. } CF_3COOH) - 78.5 \text{ p.p.m.} \tag{7.3}$$

7.3 Phosphorus[2]

Although the isotope ^{31}P has 100% abundance its inherent sensitivity to n.m.r. detection is only 6.4% of that of the proton. Until recently it has been common to use large non-spinning samples with attendant loss of resolution. With modern instruments it is possible to obtain good signal-to-noise ratios with ordinary 5-mm outer diameter (o.d.) spinning tubes and conventional

---

* Double resonance techniques for determining n.m.r. parameters of other nuclei from the ^{1}H spectrum are described in Chapter 6.

† δ values are +ve for shifts to low field (see p. 63).

high resolution techniques. It may be desirable to use larger (up to 13-mm o.d.) sample tubes for dilute solutions of phosphorus compounds, and since chemical shift differences and coupling constants are often large, relatively broad lines may be tolerated. It is possible to obtain spinners for 9-mm tubes, although it may be necessary to insert a plastic plug above the sample to prevent vortexing. Figure 7.1 gives an example of a high resolution ^{31}P spectrum.

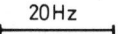

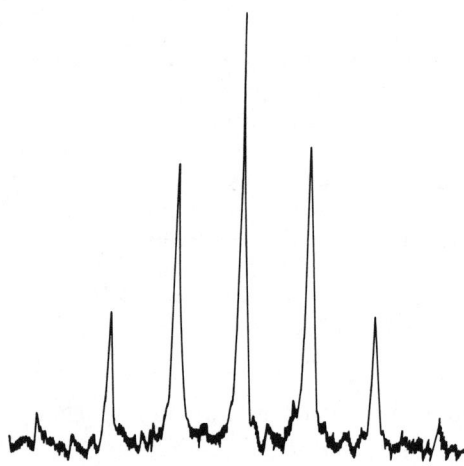

Figure 7.1 ^{31}P n.m.r. spectrum of dimethyl phosphite $(CH_3O)_2P(O)H$. The figure shows one of the two septets that form the ^{31}P spectrum of dimethyl phosphite. A doublet splitting of 690 Hz is caused by the hydrogen directly bonded to the phosphorus atom. Coupling between the phosphorus atom and the six equivalent hydrogen atoms of the methyl groups split each component of the doublet into a septet with

$$^3J_{POCH} = 12 \text{ Hz}$$

The most commonly used reference compound for phosphorus n.m.r. spectroscopy is 85% phosphoric acid which gives a single rather broad line; this is always used as an external reference. An alternative reference is P_4O_6 (external reference); this compound is liquid at ordinary temperatures and it gives a very sharp, intense n.m.r. absorption. Unfortunately this compound has the disadvantage that it is very expensive and must be kept in a sealed tube. The resonance from P_4O_6 occurs at 112.5 ± 0.1 p.p.m. to low field of that from phosphoric acid. Calibration of phosphorus spectra may be done by the usual audio-frequency sideband technique; it is often convenient to observe the relevant portion of the spectrum on the oscilloscope and to alter the audio frequency until a sideband of the reference peak and the line from the experimental sample are superposed. For samples giving weak signals it is best to use a substitution technique for calibration: the line whose position is to be measured is centred on the oscilloscope and the sample is quickly replaced by a tube containing the reference compound. The frequency of the audio oscillator is adjusted until one of the modulation sidebands is similarly centred on the oscilloscope. Repetition of the whole process will reduce the effects of magnetic field drift. Spectrometers are now available in which the field may be locked to a signal derived either from protons in the sample (ideally from tetramethylsilane) or from protons in an external water sample, while the resonance of another nuclear species is observed. It is not normally necessary to calibrate each separate spectrum obtained on such an instrument.

Solvents can have a moderate effect upon ^{31}P chemical shifts, and it is seldom worthwhile determining these to closer than ± 0.5 p.p.m. unless detailed solvent studies are being undertaken. Similarly bulk magnetic susceptibility corrections are generally ignored except in the most precise work.

7.4 Boron[3]

The isotope ^{11}B has an abundance of 81.2% and a sensitivity 16.5% that of the proton. It has a spin quantum number of $\frac{3}{2}$, and although the nuclear quadrupole moment is small the lines are normally some tens of hertz in width. Exceptions arise when the boron atom is at a site of high symmetry as in the ion BH_4^-; in these circumstances sharp lines are obtained. Interest is generally confined to chemical shifts rather than coupling constants, consequently there are usually no disadvantages attached to the use of large non-spinning sample tubes. External reference compounds are usually used for the calibration of boron spectra, boron trifluoride diethyletherate BF_3Et_2O, being the preferred reference compound. Approximate relations between three reference scales for boron compounds are given in Equations (7.4) to (7.6).

$$\delta[\text{rel. } B(OCH_3)_3] = \delta(\text{rel. } BF_3Et_2O) - 18.3 \text{ p.p.m.} \qquad (7.4)$$

$$\delta[\text{rel. } B(OCH_3)_3] = \delta(\text{rel. } BCl_3) \quad + 29.6 \text{ p.p.m.} \qquad (7.5)$$

$$\delta(\text{rel. } BCl_3) \quad = \delta(\text{rel. } BF_3Et_2O) - 47.0 \text{ p.p.m.} \qquad (7.6)$$

For boron hydride derivatives splittings arising from the boron hydrogen spin coupling interaction can frequently be resolved; the magnitudes of the coupling constant can be used to distinguish terminal ($J_{B-H} \sim 150$ Hz) and bridging ($J_{B-H} \sim 40$ Hz) hydrogen atoms.

The other naturally occurring boron isotope, ^{10}B, has an abundance of 18.8%, and only 2% of the sensitivity to n.m.r. detection of the proton. It also has a quadrupole moment which leads to broad lines, and its resonance is seldom observed in chemical studies as the same information can be obtained by using ^{11}B.

7.5 Carbon[4]

The isotope ^{13}C has an abundance of 1.1%, a spin of $\frac{1}{2}$ and a sensitivity to n.m.r. detection which is 1.6% of that of the proton. The common isotope ^{12}C has no spin. However, owing to the importance of carbon to organic chemistry much work has been done using the ^{13}C resonance, although until recently such studies have been attended by considerable experimental difficulties. Much of the earlier work was done with isotopically enriched samples, containing 50–60% ^{13}C, and spectra were often recorded under conditions of adiabatic fast passage. This technique uses very high r.f. field intensities to maximise the signal strength, together with rapid sweep rates to avoid saturation. The spectra were normally recorded in the derivative mode and great accuracy was not attained in the measurement of line positions. Recently, however, two important instrumental developments have completely altered this situation, and it is now possible to obtain high-quality spectra routinely from samples which contain ^{13}C in natural abundance. The two developments are the use of proton wide-band noise decoupling and Fourier transform spectroscopy, and they may be used either singly or in conjunction.

The ^{13}C spectra of organic molecules are normally complicated by spin coupling to directly bound protons ($J \sim 140$ Hz) and more remote protons ($J \sim 1$ to 10 Hz), and the available intensity is therefore distributed over a large number of very weak lines. This fine structure can be removed by simultaneous irradiation at the proton-resonant frequencies with an r.f. field that is sufficiently intense to give complete decoupling, and each chemically distinct ^{13}C nucleus should then give a single sharp line as its resonance. Homonuclear $^{13}C-^{13}C$ spin coupling is unimportant in samples which are not isotopically enriched, because the natural abundance of multiply substituted molecules is so low. Technically it may be difficult to achieve essentially *complete* decoupling if a single proton irradiating frequency is used, because of the large couplings involved, and the spread of the proton spectrum which may be 10 p.p.m. (corresponding to 1000 Hz at a field strength of 2.3 tesla). However, if the proton decoupling radio frequency is simultaneously modulated (see p. 78) with audio-frequency random noise, this will spread out the available power over a band of controllable width and permit much more effective decoupling. Several manufacturers provide accessories to do this and normally decoupling can be achieved over a range of several kHz with a single setting of the decoupling frequency. It is generally convenient to record the ^{13}C spectrum in the frequency-sweep mode of operation, as this avoids the need to adjust the proton frequency as the spectrum (which may extend over several hundred p.p.m.) is traversed. However, one manufacturer has developed a system in which this is done automatically, and so gives high-quality field-sweep proton-decoupled spectra.

In addition to the gain in signal-to-noise ratio which results from the concentration of all the signal intensity for a particular ^{13}C nucleus in a single line, there is also a substantial contribution from the nuclear Overhauser effect (p. 97). If the relaxation of the observed nuclear spins is dominated by *dipolar coupling** to the spins of the nuclei which are irradiated, then the factor for the enhancement of signal intensity resulting from this effect can be as great as $1 + \frac{1}{2}(\gamma_X/\gamma_A)$ where γ_A and γ_X are the magnetogyric ratios of the observed and irradiated nuclei respectively.

This factor is about three for ^{13}C proton-decoupled spectra, and enhancements close to this are commonly achieved when the carbon atom in question has one or more directly attached protons which normally dominate the relaxation. The overall gain in signal intensity that is achieved by proton noise decoupling is normally so great that samples containing up to about ten carbon atoms per molecule can give good natural abundance ^{13}C spectra without any additional enhancement being necessary. It is common practice to use 8-mm spinning sample tubes, and the slight loss of resolution that this imparts (line-widths may be up to 1 Hz) is seldom important.

For larger molecules spectral accumulation with a c.a.t. or similar device (p. 79) may be necessary, and if more than a few dozen scans are needed to give an adequate signal-to-noise ratio the process becomes extremely time-consuming. The technique of Fourier transform spectroscopy removes this difficulty. If the sample is subject to a very short intense pulse of radio-frequency power close to the ^{13}C resonant frequency, the entire spin system will be seriously perturbed and the resultant exponential decay pattern as it regains equilibrium will contain all the information present in a conventional slow passage spectrum. This decay pattern may be recorded in a few seconds rather than perhaps five minutes, and then converted into a normal spectrum by means of a Fourier transformation program in a small dedicated computer. A large number of decay patterns may thus be accumulated at intervals of say ten seconds in a c.a.t. in a relatively short time, and the result can be finally transformed into a normal spectrum of high signal-to-noise ratio. With equipment that is currently available there is no difficulty in accumulating the decay patterns from many thousands of pulses, and thus obtaining high-quality ^{13}C spectra from extremely large molecules in quite short periods of time. The time required can be further reduced by increasing the pulse rate to perhaps one per second, so that the system relaxes to only a small extent between pulses and is maintained in a state of dynamic equilibrium. This technique is known as Driven Equilibrium Fourier Transform (DEFT) spectroscopy, and the pulse sequence needed for this and other variations is normally controlled by the same small computer that is used to accumulate the decay patterns and effect the Fourier transformation.

Figure 7.2 (b) shows the ^{13}C proton-decoupled spectrum of isoquinoline obtained by the conventional continuous wave technique, and it is clear that the problem of assignment is much simpler than for the highly complex proton spectrum of the same molecule shown in Figure 7.2 (a). Figure 7.3 (a) shows the decay envelope acquired from 192 pulses of a sample of α- and β-D-Glucose, and Figure 7.3 (b) shows the corresponding Fourier transformed spectrum.

Clearly the use of proton decoupling in the study of ^{13}C spectra involves a considerable loss of information, and consequently the problems of assignment can be formidable. Fortunately several techniques are available which

---

* This is completely analogous to the direct magnetic interaction between a pair of bar magnets, and is independent of chemical bonding.

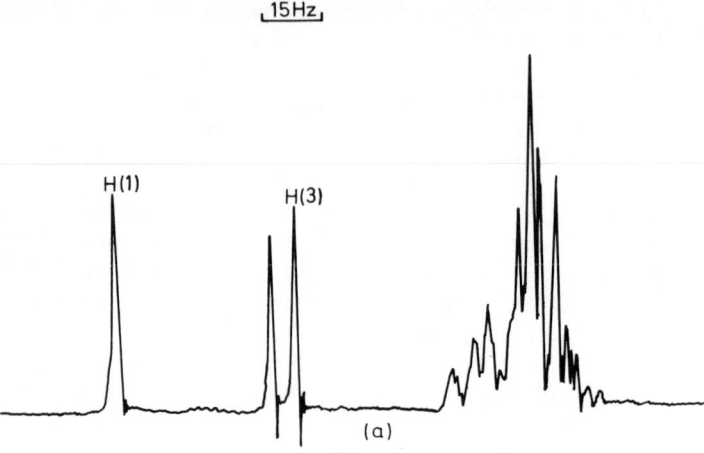

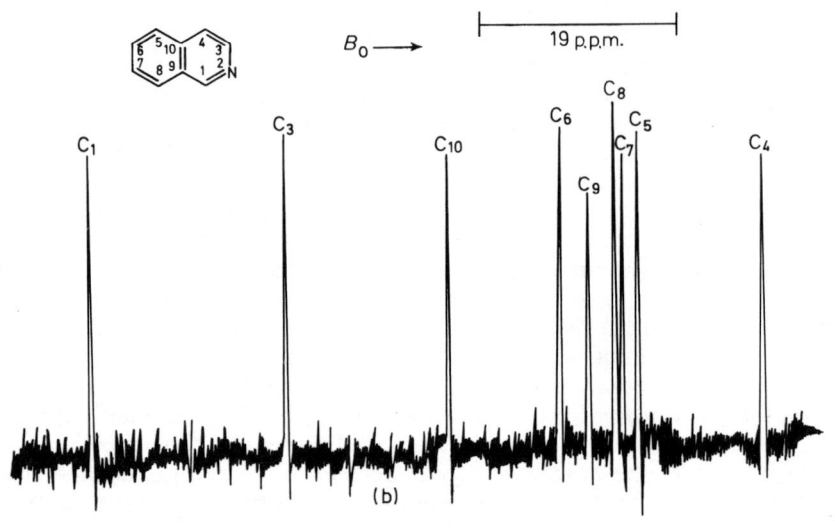

Figure 7.2.

(a) The ¹H n.m.r. spectrum of isoquinoline. The peaks arising from H(1) and H(3) can be assigned, as these have chemical shifts that are considerably different from the other protons. The remainder of the spectrum consists of an incompletely resolved group of multiplets arising from the remaining five protons; assignment of the peaks is not straightforward.

(b) ¹³C (proton decoupled) n.m.r. spectrum of isoquinoline. The spectrum consists of nine single lines, one from each of the carbon sites in the molecule. For details of the assignment of these peaks see Pugmire, R. J., Grant, D. M., Robins, M. J., and Robins, R. K., *J. Amer. Chem. Soc.*, **91**, 6381 (1969). (Courtesy of Japanese Electron Optics Laboratory)

alleviate this difficulty, and these include incomplete or off-resonance decoupling and the study of relaxation effects.

If the proton decoupling power used is barely sufficient and/or the random noise bandwith is quite narrow, it is necessary to set the proton frequency quite accurately to achieve complete decoupling. If this is not done the ^{13}C spectrum will show a partial return of the coupling to the directly bound protons and each line will become a *narrow* quartet, triplet, or doublet, according to whether it arises from a primary, secondary, or tertiary carbon

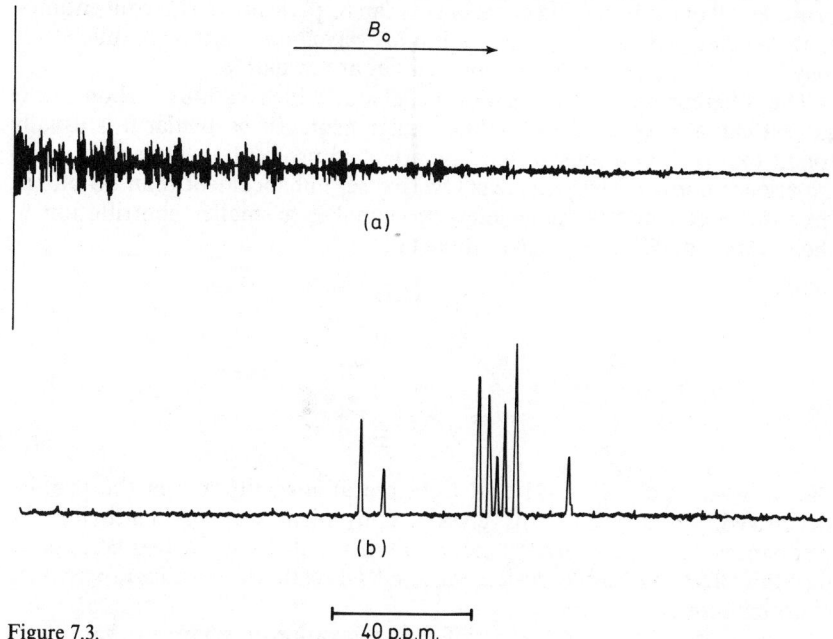

Figure 7.3.

40 p.p.m.

(a) Exponential decay envelope accumulated from 192 25-MHz pulses of a mixture of α- and β-D-Glucose in D_2O, with protons decoupled.
(b) Fourier transform of above pattern giving a conventional high resolution ^{13}C spectrum. (Courtesy of Japanese Electron Optics Laboratory)

atom. Thus by reducing the decoupling power somewhat, or by off-setting the frequency by *ca* 1 kHz, it is possible to remove the long range ^1H–^{13}C couplings (of up to say 10 Hz) and to retain the nuclear Overhauser effect enhancement, without losing the information about the number of hydrogen atoms directly attached to a particular carbon. The deterioration in signal-to-noise ratio when this is done is not very serious, and in particular quaternary carbon atoms may be readily identified since their resonance suffers no loss of intensity as the decoupling field is offset from exact resonance.

A more sophisticated use of the off-resonance technique is to determine the precise frequency (to within a few Hz) which effects optimum decoupling for a particular carbon site. This frequency then gives the chemical shift of the associated protons, and thus shows whether the carbon atom is aliphatic, olefinic, aromatic, etc. The addition of a paramagnetic *shift reagent* to the sample will affect the proton shielding by several p.p.m. (p. 30), and may alter the ^{13}C chemical shifts by up to ten times this amount. For molecules in which electron delocalisation is absent, the relative extent of the paramagnetic shift can be estimated on geometrical considerations, and for a particular carbon atom it should parallel that found for the associated protons. However, it must be stressed that this can only be regarded as an approximate guide to assignment. A further advantage of using shift reagents is that the spread achieved in the *proton* spectrum permits much more selective double resonance experiments to be performed, and it is possible for example to determine which carbon resonances are associated with *particular* aromatic protons. All of the foregoing experiments can be performed with conventional spectrometers operating in the continuous wave mode, although much time may be saved if Fourier transform facilities are available.

The different types of relaxation behaviour which various carbon nuclei exhibit can also be used for making assignments. In particular it is usually found that carbon atoms without an attached proton (i.e. quaternary ones) experience a much lower nuclear Overhauser enhancement than do others, because direct dipole-dipole interaction makes a smaller contribution to their relaxation. Thus in acenaphthylene

the resonances due to C-11 and C-12 are abnormally low in the proton-decoupled ^{13}C spectra. This gives a convenient way of identifying the resonances due to quaternary carbon atoms, but it should also be realised that this effect can lead to misleading results if relative intensities are used for counting atoms.

Another consequence of the different relaxation mechanism that affects the nuclei of quaternary carbon atoms is that the longitudinal relaxation time, T_1, is normally much longer than for other ^{13}C nuclei. This means that the pulse repetition interval in Fourier transform experiments must be greater if essentially complete relaxation is to be achieved, and the way in which the signal intensity associated with quaternary carbon atoms varies as the pulse interval is changed can be used to identify these resonances. Figure 7.4 shows the ^{13}C proton-decoupled Fourier transform spectra of quinoline

obtained with pulse repetition intervals of 7.5, 15, and 30 s. Whilst the intensities of most of the lines remain constant, the resonances given by the quaternary carbon atoms C-9 and C-10 increase in intensity as more time is provided for relaxation. The Fourier transform technique also makes it easy, by applying the appropriate pulse sequence, to determine the longitudinal relaxation times of individual ^{13}C nuclei. These often depend upon the motional freedom of the atom in question and may vary considerably in a lengthy hydrocarbon chain attached, for example, to a fatty acid residue. The study of ^{13}C relaxation times is of importance in systems of biological interest.

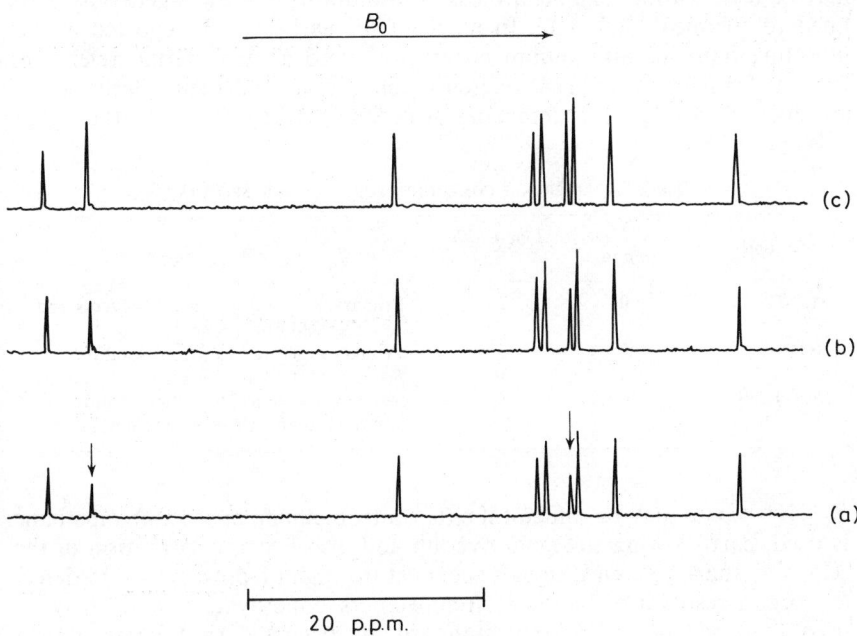

Figure 7.4. Fourier transform 25-MHz proton-decoupled ^{13}C spectra of quinoline with different pulse repetition intervals. (a) Pulse interval 7.5 s. The two vertical arrows indicate the resonances due to the quaternary carbons C–9 and C–10, which are of lower intensity because incomplete relaxation has taken place between pulses; (b) pulse interval 15 s. The resonances of C–9 and C–10 are stronger; (c) pulse interval 30 s. Relaxation is now essentially complete between pulses, and all lines are of similar height. (Courtesy of Japanese Electron Optics Laboratory)

A number of different reference compounds have been used for ^{13}C work in the past, but now that the need for an intense signal no longer exists it seems likely that internal tetramethylsilane containing ^{13}C in natural abundance will be adopted. This gives a resonance to high field of most other organic compounds, and the scale is related to those for other reference compounds by Equations (7.7) to (7.9). These equations apply when the convention adopted is that resonances at low field of the reference are given positive chemical shifts.

$$\delta(\text{rel. Si(CH}_3)_4) = \delta(\text{rel. CS}_2) \qquad +193.0 \qquad (7.7)$$

$$\delta(\text{rel. Si(CH}_3)_4) = \delta(\text{rel. C}_6\text{H}_6) \qquad +128.0 \qquad (7.8)$$

$$\delta(\text{rel. Si(CH}_3)_4) = \delta(\text{rel. CH}_3\text{COOH}) +178 \qquad (7.9)$$

7.6 Nitrogen[5]

The isotope ^{14}N has a spin quantum number of 1, an abundance of 99.6%, and only 0.1% of the proton's sensitivity to n.m.r. detection. The nuclear quadrupole moment leads to rather broad lines so that large sample tubes may be used without significant loss of resolution; interest is generally confined to chemical shift data. In much of the earlier work reported in the literature aqueous ammonium nitrate was used as an external reference. Recently[6] suggested internal reference compounds and their relative shifts are given in Table 7.1 Referencing of broad ^{14}N resonances in the range

Table 7.1 REFERENCE COMPOUNDS FOR ^{14}N N.M.R. SPECTRA

Compound	(^{14}N) p.p.m.	Line-width Hz	Uses
MeNO$_2$	0	24	Primary standard in organic solvents and concentrated mineral acids
NO$_3^-$	0±1	12	Primary standard in aqueous solution
C(NO$_2$)$_4$	−48±0.5	10	Secondary standard for nitro compounds
Me$_2$NCHO	−276±1	135	Secondary standard for aromatic nitro compounds and nitrogen heterocyclics

$\delta_N \pm 100$ p.p.m. may be difficult if either nitromethane or tetranitromethane is used, partly owing to signal overlap and also because saturation of the relatively sharp reference signals occurs at the high r.f. powers used to detect the broad resonance. In these circumstances dimethylformamide may be used as a secondary internal standard, its chemical shift being almost independent of the solvent used. Some work has been done with the isotope ^{15}N, which has a spin quantum number of $\frac{1}{2}$ and consequently no nuclear quadrupole moment. However, the low abundance (0.35%) and low sensitivity to n.m.r. detection (0.1% of that of the proton) have generally necessitated the use of enriched samples. Recently proton decoupling and Fourier transform techniques have been used to give high-quality spectra from samples containing ^{15}N in natural abundance.

An interesting point in connection with proton-decoupled ^{15}N spectra is that the magnetogyric ratio of ^{15}N is negative, and this leads to a *negative* nuclear Overhauser enhancement factor of as much as -4, when the relaxation is dominated by dipole-dipole interaction with associated protons. Thus inverted signals are obtained, and at certain proton exchange rates the relative importance of the dipole-dipole relaxation may be such that the Overhauser enhancement factor is zero and the signal vanishes. Differences in chemical shifts according to which isotope of nitrogen is studied can be regarded as negligible for chemical purposes.

The n.m.r. spectra of many other elements have been studied, including ^2H, ^{17}O, ^{29}Si, ^{119}Sn, and ^{195}Pt. Many of these are inherently difficult to investigate because of low sensitivity or abundance, or unfavourable relaxation times, and often specialised techniques must be used.

References

1. MOONEY, E. F., and WINSON, P. H., *Ann. Rev. N.m.r.*, **1**, 243 (1968)
   JONES, K., and MOONEY, E. F., *Ann. Rev. N.m.r.*, **3**, 261 (1970)
2. CRUTCHFIELD, M. M., DUNGAN, C. H., LETCHER, J. H., MARK, J., and VAN WAZER, J. R., *Topics in Phosphorus Chemistry*, **5**, Interscience (1967)
   MAVEL, G., *Prog. in N.m.r. Spectro.*, **1**, 251, Pergamon (1966)
3. HENDERSON, W. G., and MOONEY, E. F., *Ann. Rev. N.m.r.*, **2**, 219 (1969)
4. MOONEY, E. F., and WINSON, P. H., *Ann. Rev. N.m.r.*, **2**, 153 (1969)
   RANDALL, E. W., *Chemistry in Britain*, **7**, 371 (1971)
5. MOONEY, E. F., and WINSON, P. H., *Ann. Rev. N.m.r.*, **2**, 125 (1969)
   HERBISON-EVANS, D., and RICHARDS, R. E., *Mol. Phys.*, **8**, 19 (1964)
   RANDALL, E. W., and GILLES, D. G., *Prog. in N.m.r.*, **6**, 119 (1971)
6. WITANOWSKI, M., and JANUSZEWSKI, H. J., *Chem. Soc.* (B), 1062 (1967)
7. SILVER, B. L., and LUZ, Z., *Q. Rev. Chem. Soc.*, **21**, 458 (1967)

Appendix 1   Nuclear properties of some isotopes

Nucleide	Natural abundance (%)	I	Magnetogyric ratio γ (radians s^{-1} T^{-1})	Relative sensitivity at constant field for equal numbers of nuclei	Resonance frequency at 1.4090 T (MHz)	Resonance frequency at 2.3010 T (MHz)	Reference or standard compound
^{1}H	99.98	$\frac{1}{2}$	2.676×10^8	1.000	60.00	100.00	$(CH_3)_4Si$
^{2}H	0.016	1	4.106×10^7	0.0097	9.21	15.35	D_2O
^{7}Li	92.57	$\frac{3}{2}$	1.040×10^8	0.294	23.31	36.43	Aqueous LiBr
^{11}B	81.17	$\frac{3}{2}$	8.582×10^7	0.165	19.25	32.08	$Et_2O \cdot BF_3$
^{13}C	1.11	$\frac{1}{2}$	6.725×10^7	0.016	15.09	25.14	$CS_2:CH_3COOH:$ $(CH_3)_4Si$
^{14}N	99.64	1	1.933×10^7	0.001	4.33	7.22	$MeNO_2:NO_3^-$
^{15}N	0.36	$\frac{1}{2}$	-2.711×10^7	0.001	6.08	10.13	
^{17}O	0.037	$\frac{5}{2}$	-3.628×10^7	0.029	8.13	12.71	
^{19}F	100.0	$\frac{1}{2}$	2.517×10^8	0.830	56.45	94.08	$CFCl_3:CF_3COOH$
^{23}Na	100.0	$\frac{3}{2}$	7.075×10^7	0.093	15.86	24.78	
^{27}Al	100.0	$\frac{5}{2}$	6.97×10^7	0.206	15.63	26.06	
^{29}Si	4.7	$\frac{1}{2}$	-5.314×10^7	0.078	11.92	19.87	
^{31}P	100.0	$\frac{1}{2}$	1.082×10^8	0.066	24.29	40.48	$H_3PO_4:P_4O_6$
^{77}Se	7.5	$\frac{1}{2}$	5.109×10^7	0.007	11.46	19.10	$H_2Se:Me_2Se$
^{103}Rh	100.0	$\frac{1}{2}$	-8.419×10^6	0.00003	1.88	3.15	
^{111}Cd	12.86	$\frac{1}{2}$	-5.672×10^7	0.0095	12.72	21.20	
^{117}Sn	7.67	$\frac{1}{2}$	-9.910×10^7	0.045	22.61	34.72	
^{119}Sn	8.68	$\frac{1}{2}$	-9.971×10^7	0.052	22.36	37.27	Me_4Sn
^{125}Te	7.03	$\frac{1}{2}$	-8.45×10^7	0.0316	18.95	31.59	
^{129}Xe	26.24	$\frac{1}{2}$	-7.43×10^7	0.0212	16.59	25.93	
^{183}W	14.28	$\frac{1}{2}$	1.099×10^7	0.00007	2.47	4.11	
^{195}Pt	33.7	$\frac{1}{2}$	5.751×10^7	0.010	12.90	21.50	
^{199}Hg	16.86	$\frac{1}{2}$	4.77×10^7	0.0057	10.71	17.85	
^{203}Tl	29.52	$\frac{1}{2}$	1.529×10^8	0.187	34.27	55.98	Aqueous $Tl(NO_3)_3$
^{205}Tl	70.5	$\frac{1}{2}$	1.54×10^8	0.192	34.62	56.54	$Pb:Me_4Pb$
^{207}Pb	21.1	$\frac{1}{2}$	5.59×10^7	0.009	12.54	20.90	

Appendix 2 Proton chemical shifts

A2.1 Chemical shifts of methyl protons (δ– values in p.p.m.)

X	MeX	$MeCH_2X$	$MeCH_2CH_2X$	Me_2CHX	Me_3CX
H	0.23	0.86	0.91	0.91	0.89
F	4.26	1.24	—	—	—
Cl	3.06	1.48	1.06	1.54	1.60
Br	2.69	1.65	1.04	1.71	1.75
I	2.15	1.86	1.05	1.88	1.95
OH	3.39	1.17	0.93	1.16	1.22
—OR	3.25	1.15	0.93	1.08	1.24
—CHO	2.17	1.13	0.98	1.12	1.07
—COCH$_3$	2.09	1.05	0.93	1.08	1.12
NH$_2$	2.47	1.10	0.93	1.01	1.08
CN	1.97	1.30	1.10	1.35	1.37
NO$_2$	4.29	1.58	1.03	1.57	—

A2.2 Chemical shifts of protons in methyl groups attached to double bonds (δ– values in p.p.m.)

Compound	δ	Compound	δ	
MeXC=CH$_2$				
		Me$_2$C=CHX	*cis* to X	*trans* to X
X = Cl	2.15			
Br	2.30	X = Br	1.75	1.75
CN	2.03	OAc	1.65	1.65
OAc	1.92	CH$_3$CO	2.06	1.86
COCl	2.03	HC≡C	1.88	1.80
CHO	2.10	H	1.70	1.70
Ph	2.14	Me	1.63	1.63

A2.3 Chemical shifts of methylene and methine protons (δ– values in p.p.m.)

X	CH_3CH_2X	$CH_3CH_2CH_2X$	$(CH_3)_2CHX$
H	0.86	1.33	1.33
Cl	3.47	1.81	4.14
Br	3.36	1.89	4.20
I	3.16	1.88	4.24
OH	3.59	1.53	3.49
—OR	3.37	1.56	3.55
—CHO	2.46	1.66	2.39
—COCH$_3$	2.47	1.55	2.54
NH$_2$	2.74	1.43	3.07
CN	2.33	1.71	2.67
NO$_2$	4.37	2.01	4.44

A2.4 Parameters for the prediction of olefinic proton chemical shifts (δ– values) in

$$\underset{X}{\overset{H}{>}}C=C\underset{Z}{\overset{Y}{<}}$$

Substituent	Position		
	X	Y	Z
—H	0	0	0
—R	0.44	−0.26	−0.29
—CH$_2$O	0.67	−0.02	−0.07
—CH$_2$Cl(Br)	0.72	0.12	0.07
—CH$_2$N<	0.66	−0.05	−0.23
—CN	0.23	0.78	0.58
[a]—C=C<	0.98	−0.04	−0.21
[a]—C=O	1.10	1.13	0.81
[a]—COOH	1.00	1.35	0.74
[a]—COOR	0.84	1.15	0.56
—Ph	1.35	0.37	−0.10
—Cl	1.00	0.19	0.03
—Br	1.04	0.40	0.55
—F	1.03	−0.89	−1.19

[a] In the absence of conjugation with the remainder of the molecule

The required δ– value is obtained by *adding* the sum of the appropriate parameters for substituents in each of the positions X, Y, and Z to 5.28.

A2.5 Additive substituent parameters for predicting proton chemical shifts (δ– values in p.p.m.) in substituted benzenes

Substituent	Ortho	Meta	Para
H	0	0	0
F	−0.30	−0.02	−0.22
Cl	0.02	−0.06	−0.04
Br	0.22	−0.13	−0.03
I	0.40	−0.26	−0.03
Ph	0.18	0.00	−0.08
Me	−0.17	−0.09	−0.18
OH	−0.50	−0.14	−0.40
OMe	−0.43	−0.09	−0.37
NH_2	−0.75	−0.24	−0.63
NMe_2	−0.60	−0.10	−0.62
NO_2	0.95	0.17	0.33
CHO	0.58	0.21	0.27
CN	0.27	0.11	0.30
COOH	0.80	0.14	0.20

The required δ– value is obtained by *adding* the sum of the appropriate parameters for each substituent to 7.27.

A2.6 Chemical shifts (δ– values) of aromatic and heteroaromatic protons

A2.7 Chemical shifts (δ– values in p.p.m.) of methyl groups attached to aromatic rings

CH₃ on benzene ring — 2.35

CH₃ on naphthalene (1-position) — 2.65

CH₃ on naphthalene (2-position) — 2.45

CH₃ on furan (3-position) — 1.94

CH₃ on furan (2-position) — 2.17

CH₃ on pyrrole (2-position) — 2.16

CH₃ on pyrrole (3-position) — 2.05

CH₃ on pyridine (2-position) — 2.55

CH₃ on pyridine (3-position) — 2.32

CH₃ on pyridine (4-position) — 2.37

CH₃ on thiophene (2-position) — 2.41

CH₃ on thiophene (3-position) — 2.21

A2.8 Chemical shifts (δ– values in p.p.m.) of olefinic protons in cyclic molecules

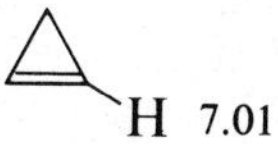

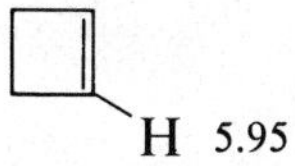

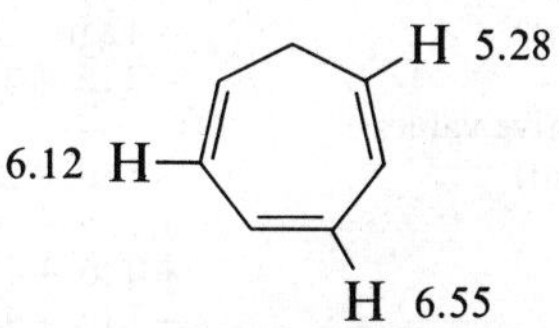

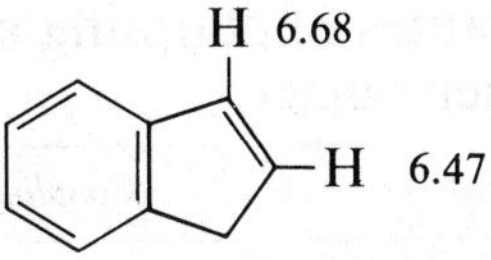

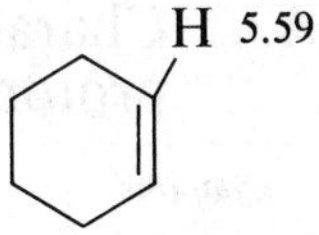

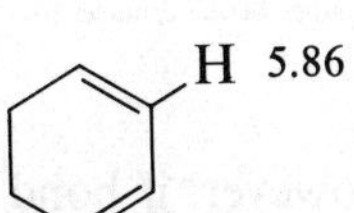

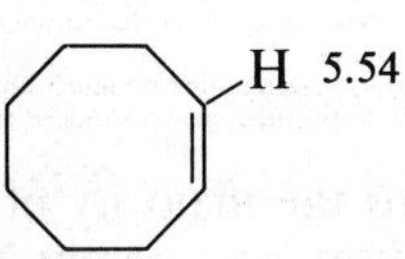

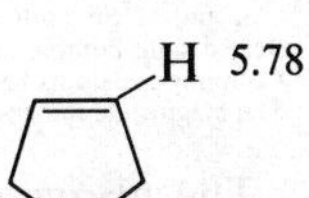

H 7.01

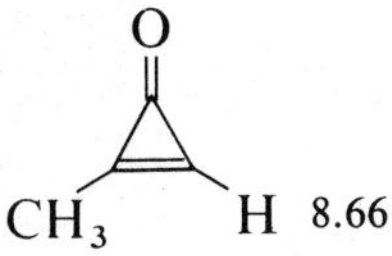

CH$_3$ H 8.66

H 5.95

H 5.28

6.12 H

H 6.55

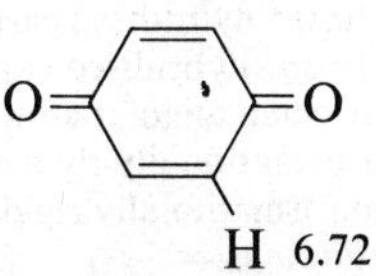

O=   =O

H 6.72

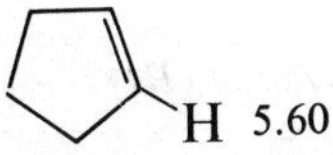

H 6.68

H 6.47

H 5.59

H 5.60

H 5.86

H 5.54

H 5.78

H 5.86

Appendix 3 Coupling constants

A3.1 Characteristic values of inter-proton coupling constants

Coupling constant	Range (Hz)
^3J(H...H) in ethyl groups with free rotation	+7 to +8
^3J(H...H)$_{cis}$ across a double bond	+5 to +12
^3J(H...H)$_{trans}$ across a double bond	+12 to +19
^2J(H...H) attached to sp^2 hybridised carbon	−3 to +3
^2J(H...H) attached to sp^3 hybridised carbon (positive values may occur when angular strain is present)	−8 to −20
^3J(H...H)$_{trans}$ in conformationally rigid systems	+7 to +14
^3J(H...H)$_{gauche}$ in conformationally rigid systems	+0 to +8
^3J(H...H) in aromatic rings	+7 to +9
^4J(H...H) in aromatic rings	+1 to +3
^5J(H...H) in aromatic rings	0 to 1

Excursions beyond the above ranges may occur when highly electro-positive substituents (e.g. Li) are present. Coupling between protons separated by more than three saturated bonds is normally less than 1 Hz, unless there is a special steric relationship (e.g. a W path).

A3.2 Characteristic values of coupling constants between protons and other nuclei

Coupling	Range (Hz)	Coupling	Range (Hz)
^1J(^{14}N–H)	+50 to +65	^2J(^{29}Si...H)	+6 to +9[c]
^1J(^{13}C–H)	+120 to +300[a]	^2J(^{119}Sn...H)	+30 to +100[c]
^1J(^{29}Si–H)	−200 to −400[bc]	^2J(^{199}Hg...H)	−100 to −220[e]
^1J(^{31}P–H)	+180 to +220 (P^{III})	^3J(^{19}F...H)	0 to +50
	+450 to +1000 (P^V)	^3J(^{31}P...H)	+10 to +30
^1J(^{11}B–H)	+80 to +200[d]	^3J(^{119}Sn...H)	−50 to −200[c]
^2J(^{19}F...H)	+40 to +80	^3J(^{199}Hg...H)	+100 to +400[e]
^2J(^{31}P...H)	−3 to +3 (P^{III})		
	−10 to −25 (P^V)		

[a] Increases as the s-character of the C–H bond and/or the electronegativity of the other groups attached to carbon increases
[b] The behaviour of silicon is similar to that of carbon as indicated in note above
[c] ^{29}Si and ^{119}Sn both have a negative magnetogyric ratio (γ) so that coupling constants involving either of these nuclei have reversed signs compared with analogous ^{13}C couplings
[d] Coupling constants between boron and *bridging* hydrogen atoms are much smaller
[e] The magnitude increases as the electronegativity of the other group attached to mercury increases

The above ranges apply to the majority of molecules; however, if bond strain or other special features are present other values of the coupling constants may be observed.

A3.3 Coupling constants of invariant sign which may be used for double resonance sign determinations

Coupling constant	Absolute sign
$^1J(^{13}C–H)$	+
$^1J(^{13}C–^{19}F)$	−
$^1J(^{31}P–H)$	+
$^1J(^{31}P–^{19}F)$	−
$^3J(H \ldots H)$ in ethyl groups with free rotation, and in aromatic rings	+

Appendix 4 Chemical shifts of nuclei other than hydrogen

Fluorine, carbon and phosphorus chemical shifts in selected compounds. In all cases a positive value indicates a resonance to low field of the reference.

A4.1 ^{19}F Chemical shifts in p.p.m. from $CCl_3 F$

$CFBr_3$	$+ \ \ 7.4$
CF_2Br_2	$+ \ \ 6.8$
CF_4	$- \ \ 63.3$
CHF_3	$- \ \ 78.6$
CH_2F_2	-143.4
CH_3F	-271.9
CF_3CCl_3	$- \ \ 82.2$
C_6H_5F	-113.1
CF_3COOH	$- \ \ 78.5$
$C_6H_5CF_3$	$- \ \ 63.9$
ClF_3	$+114.5 \ (2\,F)$ and $+2.5 \ (1\,F)$
BrF_3	$- \ \ 24.2$
PF_3	$- \ \ 36.2$
BF_3	-132.7
SF_4	$+116.5$ and 69.5
SeF_4	$+ \ \ 62.5$
BF_4^-	-149.5
SiF_4	-911.5
BrF_5	$+271.5 \ (\text{apical F}) +140.5 \ (4 \text{ basal F})$
PF_5	$- \ \ 77.8$
SF_6	$+ \ \ 48.5$
PF_6^-	$- \ \ 66.9$
$C\!-\!CF_2\!-\!C$	-136.3 to -100
$C\!-\!CF_2\!-\!N$	-130 to $- \ 87$
$C\!-\!CF_2\!-\!O$	$- \ 93$ to $- \ 72$
$CF_3\!-\!CF_2-$	$- \ 89$ to $- \ 81$
Other $CF_3\!-\!C$	$- \ 79$ to $- \ 58$
$CF_3\!-\!N$	$- \ 60$ to $- \ 42$
Substituted fluorobenzenes	-145 to $+ \ 89$

$F_{(1)}$	-105 to $- \ 72$
$F_{(2)}$	-133 to $- \ 88$
$F_{(3)}$	-206 to -145

120

A4.2 ^{31}P Chemical shifts in p.p.m. from phosphoric acid

P_4 (in CS_2)	-488
PH_3	-241
PF_3	$+ 97$
PCl_3	$+215$
PBr_3	$+222$
PI_3	$+178$
$(HO)_2P(O)H$	$+ 4.5$
$POCl_3$	$+ 5.4$
$R(R'O)_2P$	$+170$ to $+185$
$R(R'S)_2P$	$+ 69$ to $+ 85$
$(RO)_3P$	$+125$ to $+140$
$(RS)_3P$	$+ 23$ to $+ 50$
$RR'(R''O)PO$	$+ 45$ to $+ 55$
$R(R'O)_2PO$	$+ 27$ to $+ 35$
$R(R'O)(R''S)PO$	$+ 49$ to $+ 58$
$(RO)_3PO$	$- 18$ to 0
$\overset{-O}{\underset{-O}{>}}P(OR)O$	$- 15$ to $+ 19$
$(RO)_3PS$	$+ 67$ to $+ 73$
Cyclic phosphonitrilic compounds	$- 10$ to $+ 52$
PR_4^+	$+ 12$ to $+ 37$
$OPR_2(NZ_2)$ and $OPR(NZ_2)_2$	$+ 5$ to $+ 50$
$OPX_2(NZ_2)$ and $OPX(NZ_2)_2$	$- 40$ to $+ 40$
$OP(OZ)_2(NZ_2)$	$- 19$ to $+ 14$
$OP(OZ)(NZ_2)_2$ and $OP(SZ)(NZ_2)_2$	$- 4$ to $+ 21$
$OP(NZ_2)_3$	$- 16$ to $+ 40$
PRH_2	-164 to -110
PR_2H	-100 to $- 40$
PR_3	$- 65$ to $+ 7$
Ni carbonyl complex PR_3	$+ 12$ to $+ 50$
Cu complexes with $P(OR)_3$	$+100$ to $+132$

A4.3 ^{13}C Chemical shifts in p.p.m. from the ^{13}C resonance of $Si(CH_3)_4$

n-Alkanes		$+ 1$ to $+ 21$
RCH_2X		$- 2$ to $+ 72$
R_2CHX		$+ 20$ to $+ 80$
R_3CX		$+ 43$ to $+167$
Cycloalkanes		$+ 23$ to $+ 33$
Olefinic carbon		$+118$ to $+138$
Acetylenic carbon		$- 7$ to $+ 93$
$C=O$ (ketones and aldehydes)		$+193$ to $+219$
$C=O$ (carboxylic acids and esters)		$+153$ to $+183$
Methyl benzenes **C**—Me		$+133$ to $+137$
—**CH$_3$**		$+ 13$ to $+ 23$
=**CH**		$+123$ to $+133$
Mono-substituted benzenes **C**—X		$+ 96$ to $+163$
o—C		$+112$ to $+138$
m—C		$+128$ to $+131$
p—C		$+116$ to $+134$
Aromatic **C**—OH		$+153$ to $+156$
Nitriles		$+114$ to $+130$
Iso-nitriles		$+156$ to $+181$
Metal carbonyls		$+192$ to $+212$
Pyridine	α	$+149.9$
	β	$+123.8$
	γ	$+135.7$

Appendix 5

A5.1 Resonance field strengths for some nuclei at selected frequencies

Field strength T	Isotope	Frequency (MHz)
2.497	^{19}F	100
2.349	^{1}H	100
2.295	^{2}H	15
2.220	^{23}Na	25
1.830	^{11}B	25
1.511	^{7}Li	25
1.498	^{19}F	60
1.409	^{1}H	60
1.332	^{23}Na	15
1.098	^{11}B	15
1.018	^{205}Tl	25
0.907	^{7}Li	15
0.870	^{31}P	15
0.624	^{19}F	25
0.611	^{205}Tl	15
0.587	^{1}H	25
0.375	^{19}F	15
0.352	^{1}H	15

A5.2 Graph of magnetic field strength against magnet current

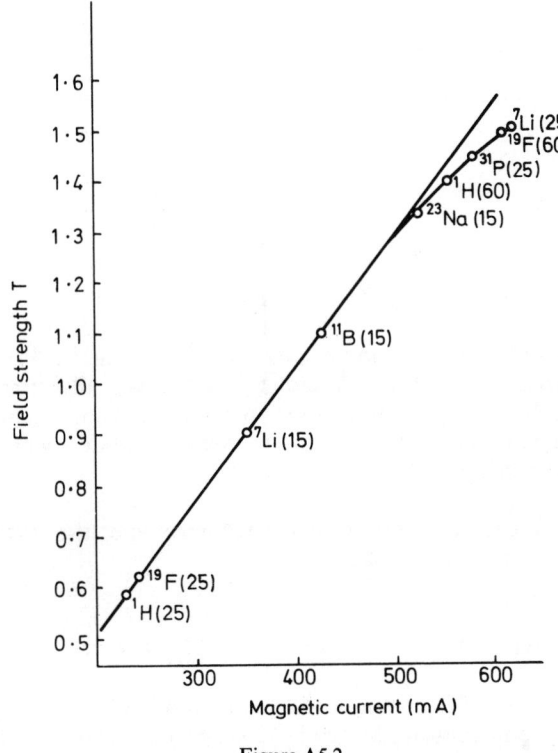

Figure A5.2.

Figures in parenthesis are resonance frequencies in MHz for the isotope.

Appendix 6   Interpreted spectra

Spectra (^{1}H unless otherwise indicated) of the following are discussed:
n-propyl alcohol; p-toluidine; pyrazole; α-D-glucose; allyl bromide; tetra-fluoroborate anion (^{19}F spectrum); phenyl thallium dichloride; 5,5-dimethyl-2-phenoxy-dioxaphosphorinane; ethyl sulphite; pentafluoroiodo-benzene (^{19}F spectrum).

These can be used as exercises if the reader studies the spectra without consulting the accompanying text.

A6.1 n-Propyl alcohol: ^{1}H spectrum at 60 MHz

In this sample the —OH protons are exchanging rapidly and so give a sharp singlet without any splitting by the α—CH$_2$ protons. The chemical shift of this singlet will be temperature and concentration dependent. The α—CH$_2$ protons give an almost first-order triplet as a result of coupling to the

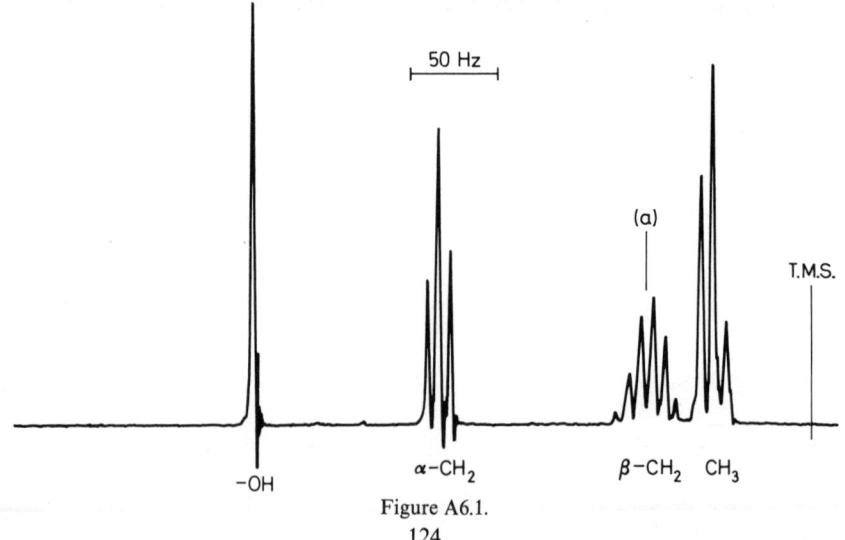

Figure A6.1.

124

β—CH$_2$ protons, and the methyl group gives a triplet in which there is severe second-order perturbation of the intensities. The β—CH$_2$ resonance should be a triplet of quartets, but since there is equal coupling to the methyl and the methylene protons there is much overlapping and only six lines are obtained. In the first-order limit the two central lines of this multiplet would be of equal intensity, but this is not so in the present system. The centre of the multiplet is indicated by the vertical line (a).

A6.2 60 MHz ^{1}H n.m.r. spectrum of *p*-toluidine (CCl$_4$ solution)

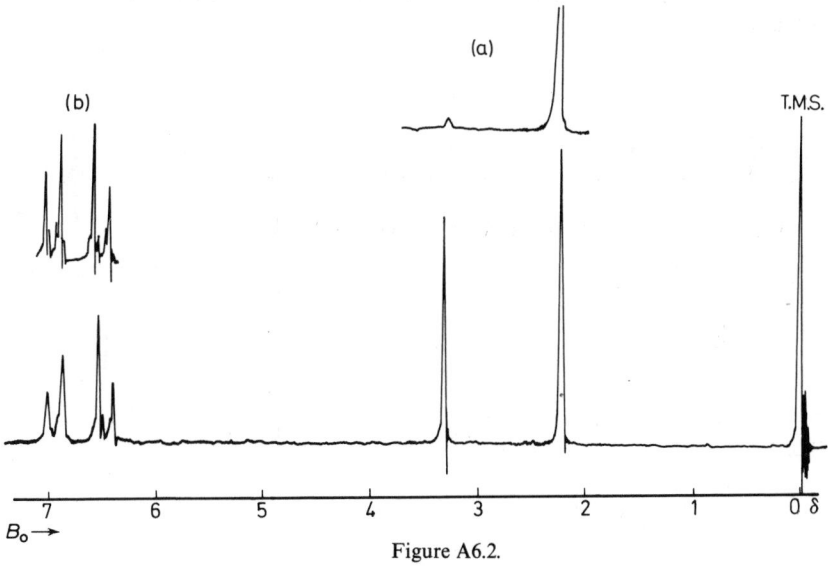

The assignment of the proton spectrum is as follows:

Singlet at $\delta = 2.23$ p.p.m.   CH$_3$
Singlet at $\delta = 3.30$ p.p.m.   NH$_2$
Multiplet at $\delta = 6.48$ p.p.m.   CH (*o*– to CH$_3$)⎱
Multiplet at $\delta = 6.95$ p.p.m.   CH (*p*– to CH$_3$)⎰ AA'BB' system

Figure A6.2.

The assignment of the peak at $\delta = 3.30$ to the protons of the NH$_2$ group may be confirmed by shaking the solution with a small amount of D$_2$O, allowing the layers to separate and then rerunning the spectrum. The labile —NH$_2$ protons exchange with the deuterium of the D$_2$O leading to a reduction in the area of the NH$_2$ peak, see inset (a).

The spectrum illustrates the dangers of estimating relative numbers of atoms from peak heights instead of peak areas. The relative *heights* of the NH_2 and CH_3 peaks are $2 : 2.7$ whereas the relative areas are $2 : 3$. The peak of the CH_3 group is broadened as a result of coupling to the ring protons giving an unresolved splitting. This is reflected in the multiplets arising from the ring protons, one of which is broader and less well resolved than the other. Homonuclear decoupling, with the second field irradiating at the frequency of the methyl group, gives a spectrum (inset (b)) for the ring protons with the multiplet at $\delta = 6.95$ much sharper than for the single resonance spectrum.

A6.3   60 MHz ^{1}H n.m.r. spectrum of pyrazole (carbon tetrachloride solution)

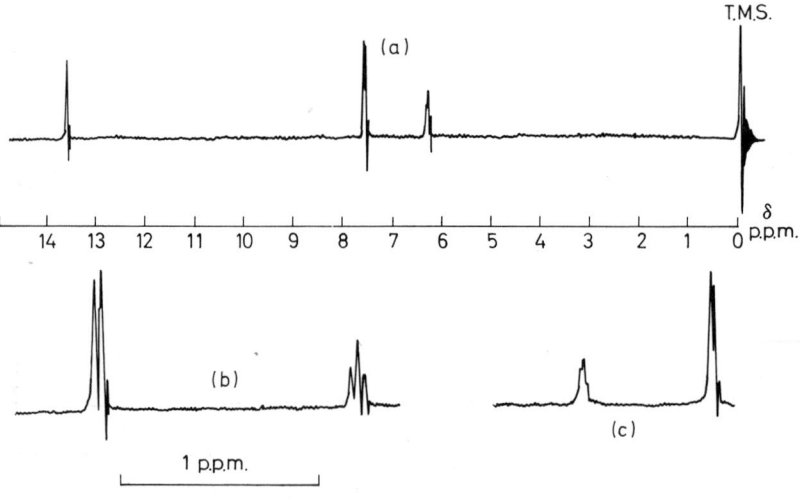

Exchange of hydrogen between N(1) and N(2) is rapid so H(3) and H(5) are equivalent.

Figure A6.3.

Figure A6.3 (a) shows the complete spectrum, from which the following assignment may be made:

Multiplet $\delta = 6.33$ H(4)
Multiplet $\delta = 7.60$ H(3) and H(5)
Singlet $\delta = 13.60$ NH

Figure A6.3 (b) The expanded record of the CH spectrum shows
$J_{H-H} = 3.8$ Hz

Figure A6.3 (c) Shows the CH spectrum of the isomeric compound imidazole.

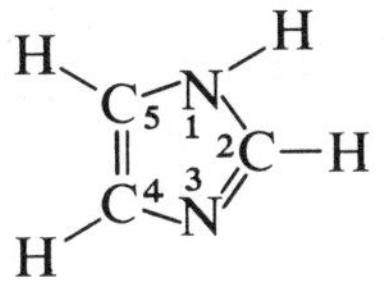

Here $J_{H-H} = 1.9$ Hz.

For this compound the CH peak of greater area is at higher field than that from the single CH, whereas for pyrazole the converse holds. The coupling constant and the relative position of CH peaks are both features which may be used to distinguish between the isomeric substituted pyrazoles and imidazoles.

A6.4 Part of the 60 MHz ^{1}H n.m.r. spectrum of α-D-glucose (DMSO solution)

Figure A6.4 illustrates the detection of different —OH resonances from a polyhydroxy compound.

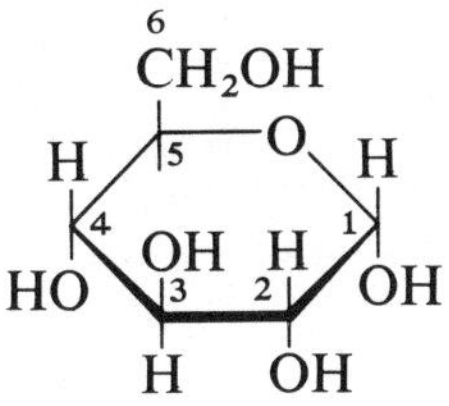

The partial spectrum given in the figure shows the multiplet peaks arising from the —OH groups and the anomeric proton H(1). H(1) and —OH(6) each give a triplet, while the other —OH groups give doublets. The inset spectrum shows the same region after the solution has been treated with D$_2$O. All —OH protons have been replaced by deuterium, so the remaining doublet must arise from H(1). The assignment of the other (high-field) triplet to —OH(6) can then be made. A. S. Perlin [*Canadian J. Chem.*, **44**, 539 (1966)] has discussed the assignment of the other —OH peaks.

A6.5 Allyl bromide

The doublet at high field arises from the CH$_2$Br protons split by H$_X$ with $J(H_X \ldots CH_2) = 6.8$ Hz. Each component of this doublet is broadened by unresolved coupling to H$_A$ and H$_B$.

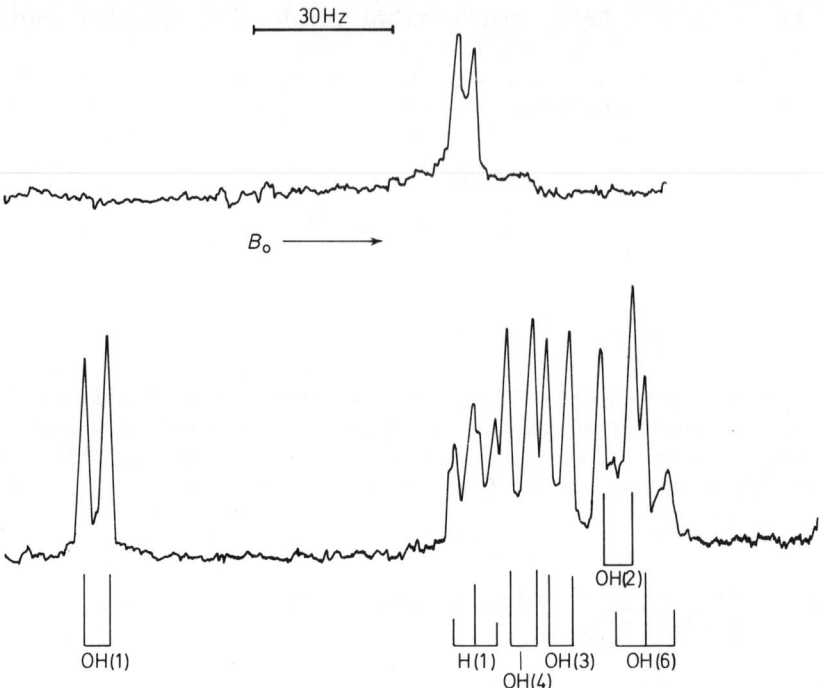

Figure A6.4.

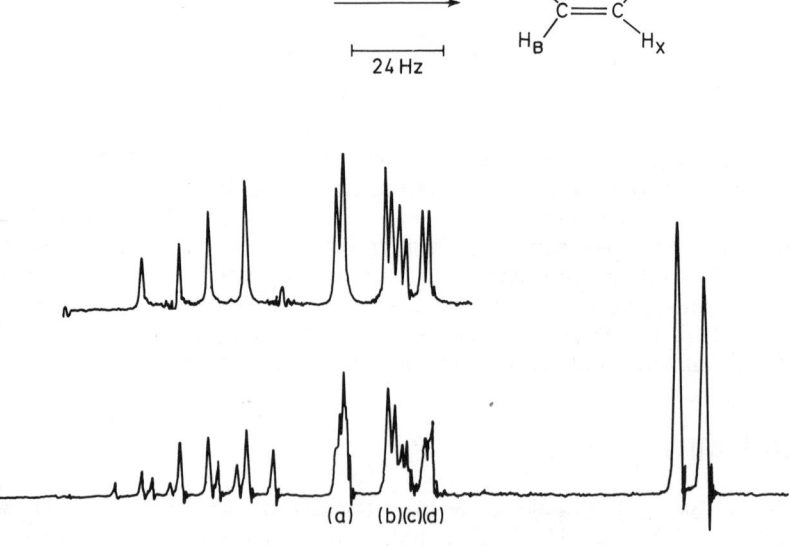

Figure A6.5.

The low-field part of the spectrum is given by the allylic protons and is complex because each line is split into a triplet by coupling to the CH_2Br protons. However, in the upper trace this complication has been removed by spin decoupling, and the spectrum is now a characteristic ABX one (*cf.* Figure 6.5 (a) Page 93) in which there is some second-order perturbation of the intensities. The four lines at lowest field are given by H_X with $J_{AX} = 17.2$ Hz and $J_{BX} = 9.8$ Hz, and this assignment is confirmed by the triplet splitting in the undecoupled trace with a spacing of $J(H_X \ldots CH_2Br) = 6.8$ Hz. H_A gives the pair of doublets labelled (a) and (c), and H_B gives (b) and (d). The narrow doublet splitting is $J_{AB} = 1.5$ Hz.

A6.6 ^{19}F spectrum of the tetrafluoroborate anion BF_4^-

Boron consists of two isotopes $^{10}B(I = 3)$ and $^{11}B(I = \frac{3}{2})$ with abundances of 18.8 and 81.2% respectively, and since each of these has a quadrupole moments there is normally much broadening in the spectra of boron-containing compounds. The symmetry of the BF_4^- anion, however, is such

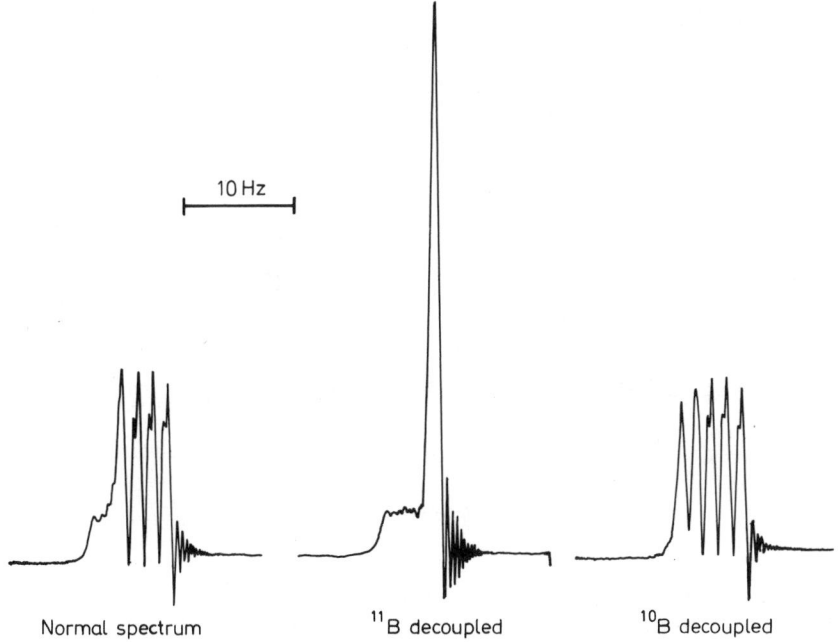

10 Hz

Normal spectrum ^{11}B decoupled ^{10}B decoupled

Figure A6.6.

that the electric field gradient at the boron nucleus is zero and so there is no quadrupole broadening in this spectrum. The strong quartet is given by ions containing ^{11}B and largely obscures the seven-line pattern from the ions containing ^{10}B. This interpretation can be confirmed by irradiating at either the ^{11}B or the ^{10}B resonance frequency to decouple the appropriate nucleus. In the first case the seven-line pattern given by the ions containing ^{10}B

becomes clearly visible, and it is also clear that there is significant isotope effect upon the ^{19}F chemical shift. This is evident in the spectrum with ^{10}B decoupled which shows that the ^{19}F shielding is *ca* 3 Hz greater in the ions containing ^{11}B.

A6.7 60 MHz ^{1}H n.m.r. spectrum of phenyl thallium dichloride (in D_6 dimethyl sulphoxide solution)

The isotopes ^{203}Tl and ^{205}Tl both have $I = \frac{1}{2}$. In the thallium aryls there is a strong spin-spin interaction between the thallium nucleus and the protons of the aromatic moiety. A further splitting occurs as a result of spin-spin interaction between the hydrogen nuclei.

The magnetogyric ratios of ^{203}Tl and ^{205}Tl are almost the same, so it is only for large couplings that the difference between $J_{203\text{Tl-H}}$ and $J_{205\text{Tl-H}}$ can be detected.

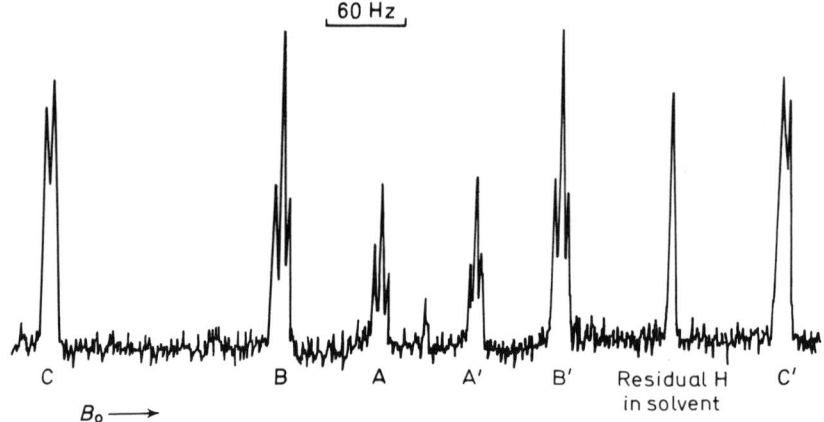

Figure A6.7.

The assignment of peaks is as follows:
AA^1 arise from the proton at the *para*-position, triplet structure arising from coupling with the two *meta*-protons. The area of A (or A^1) is half that of B (or B^1), distinguishing A and A^1 as arising from the *para*-proton with B and B^1 arising from the two *meta*-protons, the splitting here being a result of almost equal couplings $J_{o\text{H}\ m\text{H}}$ and $J_{p\text{H}-m\text{H}}$. The remaining peaks C and C^1 show a doublet splitting; these arise from the *ortho*-protons each of which is coupled to an *m*-proton. In this multiplet the thallium to hydrogen coupling is sufficiently large for the difference between $J_{203\text{Tl-H}}$ and $J_{205\text{Tl-H}}$ to be detected.

Coupling constants are:

$$J_{Tl-oH} = 852 \text{ Hz} \qquad J_{Tl-mH} = 325 \text{ Hz}$$
$$J_{Tl-pH} = 112 \text{ Hz} \qquad J_{H-H} = 9 \text{ Hz}$$

A6.8 60 MHz ^{1}H n.m.r. spectrum of 5,5-dimethyl-2-phenoxy-dioxaphosphorinane (in carbon tetrachloride solution)

The molecules of this compound preferentially adopt a chair conformation with the phenoxy group in an axial position:

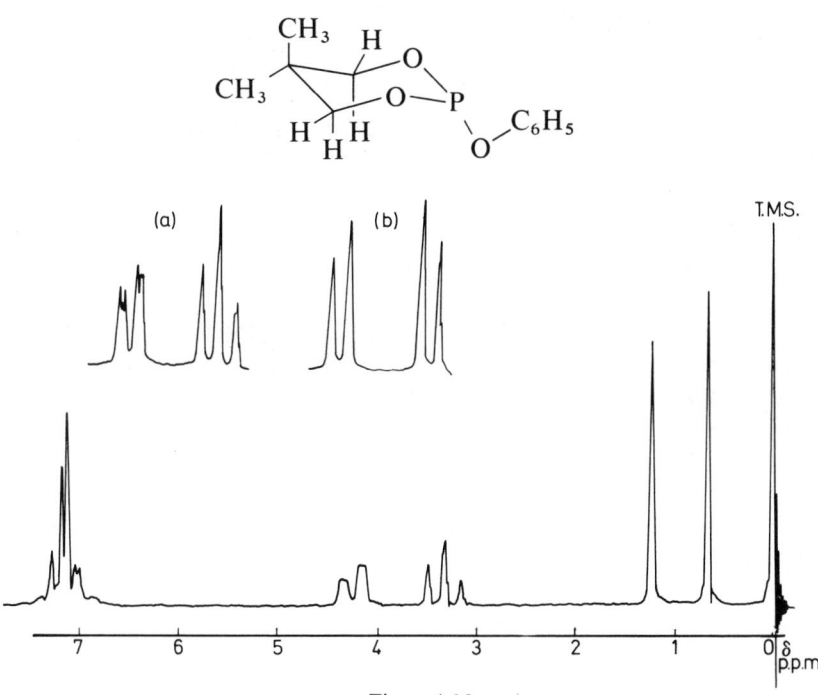

Figure A6.8.

The numbering of the ring is as follows:

$$\begin{array}{c} \text{C}-\text{O} \\ {}_6 \quad {}_1 \\ \text{C}_5 \qquad {}_2\text{P} \\ {}_4 \quad {}_3 \\ \text{C}-\text{O} \end{array}$$

The assignment of peaks is:

Singlet $\delta = 0.65$ equatorial CH$_3$ at C(5)
Singlet $\delta = 1.22$ axial CH$_3$ at C(5)
Multiplet $\delta = 3.31$ equatorial H at C(4) and C(6)
Multiplet $\delta = 4.25$ axial H at C(4) and C(6)
Multiplet $\delta = 7.15$ H of aromatic ring

The protons of the axial CH_3 are coupled to the axial protons at C(4) and C(6) giving an unresolved splitting. As a result of this coupling the axial CH_3 peak height is reduced in comparison with the height of the CH_3 equatorial peak. Inset (a) shows the spectrum obtained from the C(4) and C(6) protons when the resonance of the axial CH_3 protons is irradiated with a decoupling field. The fine structure remaining in the double resonance spectrum is due to

(i) cross ring coupling, this is an AA^1BB^1X spin system
(ii) coupling with ^{31}P.

The spectrum shows that $^{3}J_{^{31}P-H(axial)} < {}^{3}J_{^{31}P-H(equatorial)}$ the P—O—C—H dihedral angle being different for the axial and equatorial protons.

$$^{3}J_{^{31}P-H(axial)} = 2.6 \text{ Hz}, \, ^{3}J_{^{31}P-H(equatorial)} = 10.3 \text{ Hz}$$

Heteronuclear decoupling of ^{31}P (inset (b)) reduces the spectrum of the C(4) and C(6) protons to a pseudo AB multiplet which gives $^{2}J_{H(axial)-H(equatorial)} = 10.5 \text{ Hz}$.

A6.9 Ethyl sulphite: $(CH_3CH_2O)_2SO$

This pyramidal molecule might be expected to give the A_3X_2 spectrum typical of a freely rotating ethyl group, but in fact the observed pattern is more complex. This is because the *effective asymmetry* at the sulphur atom

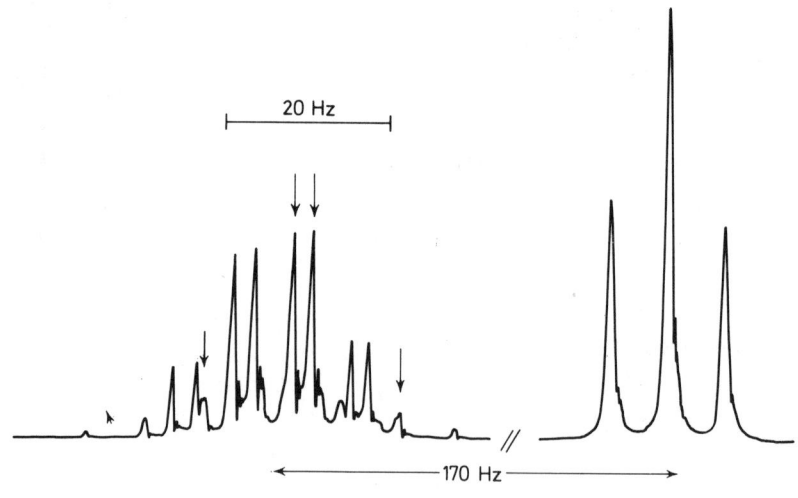

Figure A6.9.

renders the two methylene protons *within* a particular ethyl group magnetically non-equivalent. That is, from the point of view of a single methylene group the sulphur atom is a chiral centre, and internal rotations cannot lead to equivalence of shielding of the two methylene protons. Thus the spectrum must be analysed as an ABX_3 one in which coupling between H_A and H_B

has to be taken into account. In the present case the chemical shift difference between the methyl and methylene protons is large enough for a first-order approximation to be used, and so there are effectively four separate AB patterns—one for each of the possible spin arrangements of the three methyl protons. The positions of the components of one of these AB quartets are indicated by asterisks in the diagram, and yield $\delta v_{AB} = 3$ Hz and $J_{AB} = 10.8$ Hz.

A6.10 ^{19}F spectrum of pentafluoroiodobenzene, C_6F_5I

The resonance of the *para*– ^{19}F nucleus is basically a triplet of triplets as a result of coupling to F_m ($J = 19.4$ Hz) and F_o ($J = 2.0$ Hz), but an element of second-order character leads to additional splitting of the centre triplet.

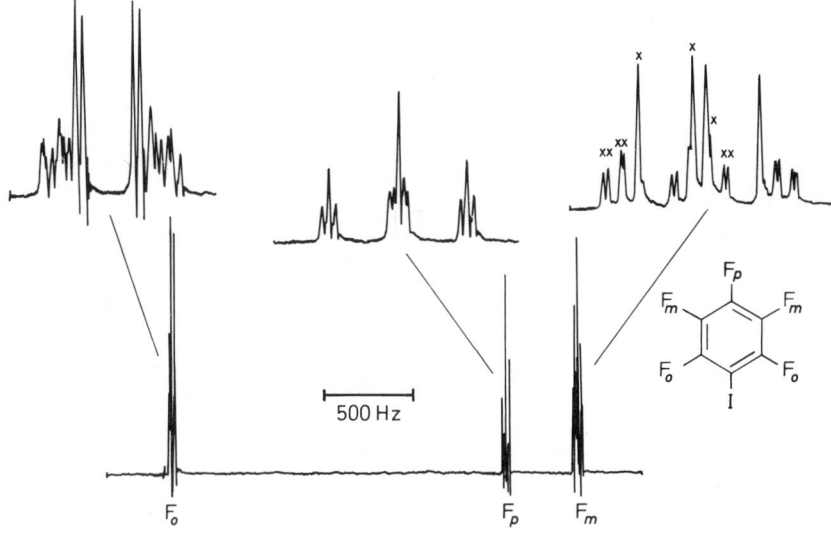

Figure A6.10.

The *meta*– ^{19}F resonance consists of two overlapping AA′XX′ patterns associated with the opposite spin states of F_p, and the lines of one of these are marked by crosses. Corresponding lines in the two AA′XX′ sub-spectra are separated by $J(F_p\ldots F_m)$. The resonance of the *ortho*– ^{19}F nuclei is similarly two overlapping AA′XX′ patterns, but these are separated by only 2 Hz. Analysis of any of these AA′XX′ spectra by standard methods would give values for $J(F_o\ldots F_m)$, $J(F_o\ldots F'_m)$, $J(F_o\ldots F'_o)$, and $J(F_m\ldots F'_m)$.

Index